T0140650

Schriftenreihe der Johanna Quandt Young Academy

Herausgegeben von der
JOHANNA QUANDT YOUNG ACADEMY AT GOETHE

Band 1

SCIENCE BETWEEN NORMS AND NATURE

Tenets, Variations, Transformations in Scientific Research

Edited by
Matthias Lutz-Bachmann, Enrico Schleiff
and Elena Wiederhold

With contributions from Sandra Eckert, Hartmut Leppin,
Matthias Lutz-Bachmann, Jürgen Mittelstraß,
Muriel Moser, and others

Franz Steiner Verlag

Bibliografische Information der Deutschen Nationalbibliothek:
Die Deutsche Nationalbibliothek verzeichnet diese Publikation in der Deutschen
Nationalbibliografie; detaillierte bibliografische Daten sind im Internet über
dnb.d-nb.de abrufbar.

© Matthias Lutz-Bachmann, Enrico Schleiff und Elena Wiederhold 2023
Veröffentlicht im Franz Steiner Verlag
www.steiner-verlag.de

Layout und Herstellung durch den Verlag
Druck: Beltz Grafische Betriebe, Bad Langensalza
Gedruckt auf säurefreiem, alterungsbeständigem Papier.
Printed in Germany.
ISBN 978-3-515-13263-3 (Print)
ISBN 978-3-515-13434-7 (E-Book)
https://doi.org/10.25162/9783515134347

Content

I. Introduction

,Nature and Normativity'[*]
Reflexionen über die Begriffe ,Natur' und ,Normativität': Das erste Jahresthema der Johanna Quandt Young Academy

MATTHIAS LUTZ-BACHMANN

Für das Jahr 2018, in dem die *Johanna Quandt Young Academy (JQYA)* an der Goethe-Universität Frankfurt am Main ihre Arbeit aufnahm, wählten die beiden Gründungdirektoren der JQYA das Thema „Nature and Normativity" aus. Damit war die Idee verbunden, die für die Anlage der wissenschaftlichen Arbeit in der *Young Academy* insgesamt leitend ist, dass alle in die Akademie aufgenommenen jungen Forscherinnen und Forscher mit ihren eigenen Arbeiten an das gemeinsame Jahresthema anschließen können. Wie produktiv gerade diese Frage nach einer Klärung der Begriffe „Natur" und „Normativität" ist, zeigte sich sehr schnell im Verlauf der ersten Debatten der Fellows und Members der JQYA. Auch die erste Auslandsakademie der JQYA in Brüssel im Herbst 2019 stand unter dem Jahresthema „Nature and Normativity". Sie versammelte Wissenschaftlerinnen und Wissenschaftler mit Vertretern der Politik in Europa am Standort der Hessischen Landesvertretung inmitten des Brüsseler Europaviertels, um sich über gesellschaftlichen Implikationen und politische Schlussfolgerungen zu verständigen, die mit den Begriffen „Natur und Normativität" verbunden sind. Im Folgenden möchte ich einige ausgewählte Aspekte der allgemeinen Frage nach dem Verständnis von „Natur" und „Normativität" und deren Verhältnisbestimmung aufzeigen. Die Pointe des gewählten Themas bestand für uns nicht nur darin, die Vielfalt der Bedeutung der Konzepte von „Natur" und „Normativität" im Blick auf die unterschiedlichen wissenschaftlichen Disziplinen komparativ herauszuarbeiten, die in der Akademie vertreten sind, und diese zu vergleichen, sondern es kam uns auch darauf an, der weiterführenden Frage nachzugehen, ob es einen inneren Zusammenhang zwi-

---

[*] Der Aufsatz ist Enrico Schleiff als Dank für die gemeinsame Arbeit in der JQYA gewidmet.

schen der Rede von „der Natur" einerseits und „dem Normativen" andererseits gibt und, wenn sich ein solcher Zusammenhang bestätigen sollte, worin dieser in der Sache begründet ist.

Bereits ein erster, flüchtiger Blick in die Begriffsgeschichte belehrt uns über die Vielfalt der Bedeutungen des Begriffs der „Natur" und zeigt, wie unterschiedlich das Naturkonzept gerade auch in der Geschichte der Wissenschaften ausgelegt worden ist. Ähnlich verhält es sich mit dem Begriff der „Normativität". Die Vielfalt der Bedeutung und der Verwendung beider Begriffe muss berücksichtigt werden, wenn insbesondere nach einem möglichen Zusammenhang zwischen der Rede von „der Natur" und dem „Normativen" gefragt wird.

Wenn wir die Begriffsgeschichte beider Konzepte betrachten, so können wir als erstes feststellen, dass mit der Frage nach der „Natur" vielfach nach einer präzisen Definition eines vorliegenden Objekts oder eines bestimmten Sachverhalts gesucht wird. Die Definition einer Sache aber resultiert aus der Bestimmung derjenigen Eigenschaften, die dieses Objekt oder diesen Sachverhalt „wesentlich" bestimmen. Wurde die Frage nach der „Natur" in der antiken Philosophie und Logik[1] genau so verstanden, dann zielte sie auf die Bestimmung des „Wesens" (lat.: „essentia") oder der „Substanz" (lat.: „substantia") einer Sache. In diesem Sinne war mit der Angabe der „Natur" einer Sache in diesem Zusammenhang genau das gemeint, was eine Sache „wesentlich" ausmacht oder „substantiell" bestimmt. Das Wissen von der Natur einer Sache, ihrem Wesen galt zugleich als die Voraussetzung dafür, dass ein wissenschaftliches Wissen auf dem Weg von Beweis oder Induktion möglich ist.[2] Auch wenn wir heute in der Wissenschaft die überlieferten Termini (wie den Begriff der „Artnatur", des „Wesens" oder der „Substanz" einer Sache) aus guten Gründen nicht mehr verwenden, so gehört es doch nach wie vor zu den zentralen Aufgaben der empirisch forschenden Natur- und Sozialwissenschaften, die konstitutiven und in diesem Sinn somit „wesentlichen" Eigenschaften von noch nicht hinreichend erkannten Objekten, von dynamischen Prozessen oder von komplexen Sachverhalten möglichst präzise zu bestimmen.

Angesichts dieser bleibenden Aufgabe von Wissenschaft, die einen „Sachverhalt"[3] konstituierenden (oder „wesentlichen") Eigenschaften präzise zu bestimmen, markiert der Begriff des „Normativen" in diesem Zusammenhang zunächst nichts anderes als die Aufgabe, dass wir in der Wissenschaft auch stets klären müssen, welche Methoden und Verfahren jeweils zum Einsatz kommen sollen, um die genannte Aufgabe zu lösen, nämlich die grundlegenden Eigenschaften komplexer Sachverhalte oder einzelner

---

1 Vgl. hierzu Aristoteles, Kategorienschrift, 1a–15b.
2 Vgl. hierzu Aristoteles, Zweite Analytik, 71a 1–13.
3 Vgl. hierzu Ludwig Wittgenstein: „1.2 Die Welt zerfällt in Tatsachen. 2 Was der Fall ist, die Tatsache, ist das Bestehen von Sachverhalten. 2.01 Der Sachverhalt ist eine Verbindung von Gegenständen (Sachen, Dingen). 2.011 Es ist dem Ding *wesentlich*, der Bestandteil eines Sachverhalts sein zu können.", in: Ludwig Wittgenstein, Tractatus logico-philosophicus, Oxford 1959, deutsche Ausgabe: Frankfurt 1969, S. 11.

Objekte zu bestimmen. So können wir hier als erstes festhalten, dass die Fragen nach der „Natur" eines Erkenntnisobjekts und nach den „Regeln" oder auch „Vorschriften", denen die Forschung folgen soll, der Sache nach bis heute – wenn auch unter anderen Termini – zum Repertoire der Wissenschaft gehört, ja Wissenschaft als begründetes Wissen von der Welt auszeichnet.

Im Blick auf diejenigen Wissenschaften, die sich mit dem Handeln von uns Menschen befassen, können wir eine Beobachtung machen, die uns einer weiteren Klärung des Begriffs des „Normativen" im Kontext der Wissenschaften näher bringt. In den Handlungswissenschaften gehört das Phänomen des „Normativen" fraglos zum Objekt- oder Gegenstandsbereich der empirischen Forschungen; denn „Normen" oder präskriptive Regeln, die ein bestimmtes Handeln nicht nur beschreiben, sondern auch einfordern oder verlangen, sind de facto ein Bestandteil der sozialen Welt: In Gesellschaft und Kultur, in Politik und Recht, im öffentlichen Raum und im Privaten gibt es eine unübersichtliche Vielfalt von Regeln, Konventionen oder Vorschriften, die auf das Handeln von uns Menschen Einfluss nehmen. Die Sozial- und Kulturwissenschaften beschäftigen sich in ihren empirischen Forschungen mit den „Normen", die diese Regeln definieren, wobei sie sich ihrerseits primär um eine deskriptive Beschreibung der sozialen Fakten bemühen. Das tun sie, wenn sie in einer Feldforschung beispielsweise danach fragen, ob in einem von ihnen untersuchten Handlungsraum soziale Interaktionen vorkommen, die durch die Angabe von bestimmten Vorschriften oder Handlungsregeln erklärt und angemessen verstanden werden können oder nicht. Solange die Wissenschaften dabei aus einem rein deskriptiven Interesse heraus fragen, wie es für die empirische Sozialforschung typisch ist, verlassen sie nicht die Perspektive eines Beobachters, der gleichsam „von außen" oder aus der Sicht einer „dritten Person" auf das Phänomen von Normativität als einem Faktum in der sozialen Welt schaut. Dies ändert sich aber kategorisch, wenn die Wissenschaft nach der praktischen Bedeutung von „Normen" aus der Perspektive der Handelnden selbst heraus fragt. Wir können sie auch die Perspektive der „ersten" oder der „zweiten Person" nennen.

In dieser Perspektive entdecken wir die vollständige Bedeutung dessen, was wir eine „Norm" nennen; denn hier dient die Rede von der „Norm" nicht mehr allein der Erklärung von Ereignissen in einem Handlungsfeld, sondern wir nehmen den „präskriptiven" Charakter wahr, der mit einer jeden Norm konstitutiv verbunden ist. „Normen" zu verstehen heißt, ihren Aufforderungscharakter zu erkennen und zu erfahren, was es heißt, zu einem bestimmten Handeln oder Unterlassen aufgefordert zu werden. Dies erfassen wir vollständig nur aus der Perspektive der „ersten" oder „zweiten Person". Da „Normen" im Raum des menschlichen Handelns mit anderen eine motivierende und handlungsanleitende Kraft entfalten, sind wir in der Perspektive der „ersten" oder „zweiten Person" veranlasst, die Rolle eines neutralen Beobachters zu verlassen, der allein an einer deskriptiven oder erklärenden Weise des Umgangs mit dem Phänomen des „Normativen" interessiert ist. In der Perspektive der „ersten" bzw. „zweiten Person" nimmt die Frage nach der „Normativität" unweigerlich den Charakter einer Suche nach den einleuchtenden Gründen an, die uns in unserem eigenen Handeln gegenüber anderen tatsächlich verpflichten oder auch nicht. In diesem Sinn beinhaltet die Rede von „Normativität" stets einen Rekurs auf ein Konzept von Deliberation und Begründung, das über die Frage nach der „Natur" eines Sachverhalts, somit über die Suche nach der Definition der wesentlichen Eigenschaften eines Sachverhalts hinausgeht.

Die Bedeutung des Begriffs von „Natur" verändert sich, wenn in der Geschichte des Begriffs „die Natur" selbst als ein reales Ganzes, ja als ein umfassender Wirkungszusammenhang verstanden wird, wie dies bei den ionischen Naturphilosophen in der griechischen Philosophie geschehen ist. Die „Natur" kann also ihrerseits auch in den Wissenschaften als ein systemisches Ganzes betrachtet werden. Dann ist „das Natürliche" tendenziell identisch mit „allem, was wird und vergeht". Mit diesem Begriff von „Natur" wird zugleich das integrative Prinzip allen „Werdens" und „Vergehens" bestimmt, das wir dann als „die Natur" beschreiben. In der Reflexion der antiken Naturphilosophen wird stets auf die etymologische Verbindung des Nomens „Physis" („Natur") mit dem Verb „phyestai" („werden") verwiesen.[4] In den modernen Biowissenschaften ist der Referenzpunkt, der mit der Angabe „der Natur" verbunden ist, alles Organische und Lebendige, das – zumindest auf unserem Planeten Erde – in einem großen „Stoffwechsel" miteinander steht; dieses untersucht die Evolutionsbiologie mit ihren Methoden auf die der „Natur" immanente Entwicklungsdynamik hin, während die Physik oder die Anorganische Chemie mit ihrer Referenz auf „die Natur" auch die sog. „unbelebte Natur" in ihre Betrachtung mit einschließen. Der Begriff „der Natur", den wir in diesem Kontext antreffen, konvergiert mit dem Untersuchungsgegenstand „alles dessen", was in einem strikten Sinn verstanden „unter den Gesetzen der Natur" steht, die als notwendig, unveränderlich und determinierend gelten. Im Rekurs auf diese Gesetze können die Phänomene „der Natur" von uns modelliert, auf diesem Weg erklärt und in diesem Sinn auch „erkannt" werden. Die „Gesetze der Natur" nehmen hier die Rolle von Ordnungsregeln ein, die nicht nur „die Natur" in ihren Prozessen selbst bestimmen, sondern als „Richtlinien" den Wissenschaften den kognitiven Weg zu einem sachgemäßen Forschen weisen. Die antiken Naturphilosophen sprechen im Blick auf die „Gesetze" der Natur auch von den „Nomoi", die „gesetzt", also vorgegeben sind und die Ordnung des „Kosmos" bestimmen. Offen bleibt bei den griechischen Naturphilosophen, wer oder was diese „Gesetze" der Natur „gesetzt" hat; offenkundig aber unterscheiden sich diese „Gesetze" der Natur von den „Gesetzen" der (griechischen) Politik. Darauf hatten nicht zuletzt die antiken Sophisten hingewiesen[5], auf die die begriffliche Unterscheidung von „Natur" und „Gesetz" zurückgeht. In dieser Unterscheidung spiegelt sich für uns eine grundlegende Differenz der Rede von „Natur" und „Normativität", insofern wir voraussetzen, dass das, was wir mit „Natur"

---

4 Vgl. hierzu etwa die Darstellung von Aristoteles in Buch A seiner „Metaphysik", 983 b 6 ff.
5 Vgl. hierzu die Darstellung des Platon in seinem Dialog „Kratylos", 383 a sowie 384 d.

bezeichnen, etwas bedeutet, das vom Menschen gänzlich unabhängig, ja ihm vorgege-ben oder auch vorgeordnet ist. Doch unsere weiteren Reflexionen werden aufzeigen, dass ein solcher Naturbegriff als unterkomplex erscheint. Falls diese Beschreibung aber zutrifft, muss das Verhältnis von „Natur" und „Normativität" anders bestimmt werden als es die antiken Naturphilosophen und die Sophistik vorgeschlagen hatten.

Diese Bedeutung der Begriffe „Natur" und regelbasierter „Normativität" der wis-senschaftlichen Erkenntnis ist noch heute in der gebräuchlichen Rede von den „Na-turwissenschaften", insb. in der Bezeichnung des Fachs der „Physik" enthalten. Diese Disziplinen sind nicht nur einfach die Wissenschaften „von der Natur" im Sinn der Angabe eines umfassenden Objektbereichs, sondern sie bedienen sich in ihren Er-kenntnisverfahren zugleich auch der sog. „natürlichen Gesetze", die als invariant, als notwendig und somit als Gesetzmäßigkeiten gelten, die dem Objektbereich „Natur" gleichsam inhärent sind. Dabei bleibt in der Regel die Frage danach unbeantwortet, ob diesen Gesetzen ihrerseits noch einmal logisch betrachtet „höhere Prinzipien" und Regularitäten zugrunde liegen, ob sie in einem strengen Sinne „gesetzt" sind – wenn wir hier schon von „Gesetzen" sprechen – oder ob die Rede von Naturgesetzen ein-fach nur eine Konvention darstellt und diese Gesetze einfach nur als „gegeben", als „vorhanden" und der menschlichen Erkenntnis „vorgegeben" betrachtet werden müs-sen. Die wissenschaftliche Forschung bestünde dann darin, diese Gesetze einfach zu „entdecken", wie die Europäer den bis in das 15. Jahrhundert noch nicht bekannten Kontinent Amerika „entdeckt" hatten. Dann spiegeln diese Gesetze, wie Wittgenstein in seinem „Tractatus" schreibt, die Realität der „Welt". Sie spiegeln die Realität alles dessen, „was der Fall ist", wobei Wittgenstein mit dieser Formulierung nicht einfach die ontologische Realität „der Dinge" meint, sondern die „Gesamtheit der Tatsachen" (im bereits oben erwähnten Sinn).[6]

Einer solchen weiten Bedeutung des Begriffs „der Natur" liegt eine wichtige ter-minologische Differenz zugrunde, die in unserem Alltag, aber auch bis hinein in die Sprache der Wissenschaften bedeutsam ist. Es ist die Differenz zwischen „dem Natür-lichen", dem in der Wirklichkeit unseres Planeten oder zumindest in der uns umgeben-den sog. „natürlichen Umwelt" immer schon „Gegebenen" im Sinne eines „Vorhande-nen" oder auch objektiv „Vorgegebenen" auf der einen Seite und dem „Artifiziellen", dem „Künstlichen" oder vom Menschen gemachten, dem „Technischen" auf der an-deren Seite. Dieser begrifflich-terminologisch wichtige Unterschied begegnet uns z. B. auch in der zeitgenössischen Debatte über „natürliche Intelligenz" im Unterschied zur

---

6 Vgl. Ludwig Wittgenstein, Tractatus, S. 11: „1 Die Welt ist alles, was der Fall ist. 1.1 Die Welt ist die Gesamt-heit der Tatsachen, nicht der Dinge. 1.11 Die Welt ist durch die Tatsachen bestimmt und dadurch, dass es *alle* Tatsachen sind. 1.12 Denn, die Gesamtheit der Tatsachen bestimmt, was der Fall ist und auch, was alles nicht der Fall ist. 1.13 Die Tatsachen im logischen Raum sind die Welt."

„artifiziellen Intelligenz".[7] Im Unterschied zum Natürlichen scheint das Artifizielle nicht einfach vorgegeben zu sein, sondern es wird betrachtet als das, was wir Menschen gemacht und hergestellt haben oder was wir „an der Natur" durch mehr oder weniger geschickten Eingriff technisch verändert haben. Mit Blick auf diese Unterscheidung der „Natur" von dem Bereich der „Technik", der „Künste" (lat.: „artes") und dem „Menschengemachten" können wir sagen, dass hier die Frage nach dem „Normativen" mit der Frage nach den „Regeln" im Sinne der Normen für die Verfahren verbunden ist, die wir Menschen verwenden, wenn wir auf „die Natur", also auf das vermeintlich oder tatsächlich „auf natürliche Weise" Gegebene artifiziell einwirken, um „das Natürliche" im Sinne des „Originären" und „Ursprünglichen" auf diesem Weg entweder technisch zu verändern oder uns anzueignen und im Sinne unserer eigenen Zwecksetzung zu verwenden. Damit ist zugleich die Frage aufgeworfen, ob es eine der Natur selbst immanente Zwecksetzung, eine Naturteleologie also gibt, wie Aristoteles und nach ihm die Stoiker meinten, oder ob der Raum der Zwecke und des Zweckdenkens ganz auf den Menschen beschränkt werden muss. Im Fall einer Beschränkung der Zwecke auf den Menschen wird der Mensch als das einzige Lebewesen verstanden, das aufgrund seiner „Natur" oder seines „Wesens" seine Handlungsziele und Handlungszwecke selbst frei wählen kann, ja wählen muss. So gesehen gehört die Freiheit des Menschen nicht nur zu seiner Natur, sondern sie konstituiert und bestimmt den Menschen in einer Weise, dass wir davon sprechen können, dass der Mensch gar nicht anders als „frei" gedacht werden muss. Wir sehen, wie auch hier die Frage nach „der Natur" mit der Frage nach dem „Normativen" verschränkt ist, und dass somit beide Themen der Sache nach nicht voneinander getrennt behandelt werden können.

Aufmerksame Vertreterinnen und Vertreter der Erkenntnis- und Wissenschaftstheorie haben darauf aufmerksam gemacht, dass nicht erst bei einem effektiven, einem gleichsam „von außen" kommenden zwecksetzenden „Eingriff" des Menschen „in die Natur" das „Artifizielle", also das vom Menschen ausgehende „Gemachte", die vermeintlich „ursprüngliche", in sich selbst bewegte „Natur" verändert und die „Natur" in einer signifikanten Weise neu gestaltet wird. Zunächst müssen wir uns fragen, was es eigentlich bedeutet, wenn aus einer Sicht auf die Welt als einem großen Naturzusammenhang der Eingriff des Menschen „in die Natur" als „unnatürlich" qualifiziert oder sogar als artifiziell disqualifiziert wird. Diese Diskussion, die in den Debatten über das Verhältnis von Natur und Technik in der Gegenwart vehement geführt wird, kann ich hier nicht weiterverfolgen. Wichtig ist für unsere Fragestellung aber die folgende Beobachtung der Erkenntniskritik. Sie macht uns darauf aufmerksam, dass bereits im Akt des Erkennens von den „Objekten in der Natur" und bei der Bestimmung ihres Wesens (oder ihrer „Natur", ganz im Sinn der ersten Bedeutung des hier erläuterten

---

7 Vgl. hierzu Matthias Lutz-Bachmann, „Articificial Intelligence" and „Human Nature": What are the Philosophical Challenges of AI ?, in: Culture e Fede XXIX (2021), S. 201–215.

Begriffs) eine kognitive Aneignung des natürlicherweise Gegebenen durch uns Menschen vorliegt. Anders gesagt: Die Verfahren der Wissenschaften, die ein Wissen über die Welt generieren, indem sie die Gegenstände „in der Welt", ja auch die Welt als Ganze zu erkennen versuchen, gestalten das konstitutiv mit, was Wittgenstein „die Welt" und de „Gesamtheit der Tatsachen" genannt hatte, aus denen die Welt besteht. In der erkenntniskritisch ansetzenden modernen Philosophie wurde stets und wird mit guten Gründen immer wieder darauf hingewiesen, dass sich die „Objekte" unserer Welterkenntnis erst durch den menschlichen Zugriff auf die Welt, das heißt erst durch das „Subjekt" des Erkennens und durch die Verfahren der Wissenschaft als Erkenntnisobjekte konstituieren. Daher können wir keinesfalls davon sprechen, dass wir die „Natur der Dinge" unmittelbar entschlüsseln und erfassen, weil jeder Erkenntnisakt von uns Menschen im Alltag und in der Wissenschaft durch die Art und Weise bestimmt ist, wie wir die Phänomene der Natur betrachten, analysieren und protokollieren. Das gilt insbesondere für die Wissenschaften, die ihre Erkenntnisgegenstände auf vielfache Weise mathematisch modellieren, um sie auf der Grundlage der durch diese Verfahren allererst bestimmbaren Eigenschaften näher qualifizieren und im Blick auf zukünftige Ereignisse zu berechnen. Wenn diese Feststellung zutrifft, dann ist bereits die elementare Erkenntnis von „Sachverhalten" im Sinn der Definition von Wittgenstein als einer „Verbindung von Gegenständen (Sachen, Dingen)"[8] in der „extramentalen", also in der „realen" Welt, als ein Prozess der Erschließung und der Deutung durch den Menschen zu verstehen, als eine spezifische Art und Weise des Begreifens der Welt gemäß den Mustern des menschlichen Erkennens und der verwendeten Algorithmen, des Forschens und des Sprechens im Sinn einer allgemeinen „Logik der Forschung" (Karl Popper). Dieser „Logik der Forschung" entspricht bei Popper die Einsicht in den prinzipiellen Fallibilitätsvorbehalt gegenüber *aller* wissenschaftlichen Erkenntnis.[9] Diese kritische Einsicht bewahrt die Wissenschaften vor dem Irrtum, ihre Erkenntnis „von der Natur" objektivistisch misszuverstehen und selbst ontologisch auszudeuten. So vermeidet es eine kritische Wissenschaft, in die Falle eines erkenntnistheoretischen Naturalismus zu geraten.

Das schließt nicht aus, dass wir auf dem Weg der Wissenschaft sachgemäß erkennen, was die „Sachverhalte" unserer Erkenntnis wesentlich bestimmt und worin die dynamischen Prozesse der Natur oder der Gesellschaft und ihrer weiteren Entwicklungen tatsächlich bestehen. Damit wird lediglich offensichtlich, was wir aus allen Wissenschaften kennen, dass die in der modernen Philosophie klassische Rede von einem „Objekt" der Erkenntnis und einem „Subjekt" des Erkennens korrelative Bestimmungen zum Ausdruck bringen. Der Begriff der „Natur" bzw. der „Welt" im Sinne Wittgensteins und der Begriff der „Wissenschaft" sind intrinsisch aufeinander bezo-

---

8 Ludwig Wittgenstein, Tractatus, S. 11.
9 Vgl. Karl Popper, Logik der Forschung, 4. verb. Auflage, Tübingen 1971, insb. S. 47–96.

gen und keines dieser Konzepte ist isoliert von dem anderen vernünftig zu verstehen. Diese Einsicht macht eine scharfe und exakte Trennung zwischen „dem Natürlichen" einerseits und „dem Artifiziellen" andererseits, zwischen dem „Gegebenen" und dem von uns „Gemachten" hinfällig. Auch wenn es richtig ist, dass wir das eine vom anderen begrifflich unterscheiden können, so ist doch dieser Unterschied nicht „ab-solut", sondern bleibt bezogen auf die Konstellation eines Gegenüber von „Natur" und Erkenntnisprozess, klassisch gesprochen von Objekt und Subjekt und – das ist wichtig – auf den jeweiligen Kontext, in dem diese Unterscheidung sinnvoll ist.

Zugleich erweist es sich, dass die Rede vom „Artifiziellen", das allem menschlichen Erkennen typischerweise eingeschrieben ist, eine Voraussetzung für die Annahme ist, dass eine wissenschaftliche Erkenntnis „der Welt" (und somit auch der „Natur" in dem hier erläuterten Sinn) möglich ist. Dadurch, dass wir Menschen auch im Modus der elaboriertesten Methoden der Wissenschaften nichts anderes als unsere eigene, als unsere „natürliche Intelligenz" (oder auch besser gesagt: unseren „natürlichen Verstand und unsere „natürliche Vernunft") zur Wirkung bringen, ist es möglich und auch erforderlich, dass wir die Ergebnisse der wissenschaftlichen Forschung der Überprüfung durch andere Teilnehmer im wissenschaftlichen Diskurs aussetzen. Das wird mit dem wissenschaftstheoretischen Postulat einer argumentativen Rechtfertigung aller Aussagen der Wissenschaft zum Ausdruck gebracht. Diese Auffassung vertreten die Repräsentanten der zeitgenössischen Wissenschaftstheorie (im deutschsprachigen Diskussionszusammenhang sind hier zu nennen u. a. Jürgen Habermas,[10] Karl-Otto Apel,[11] Friedrich Kambartel,[12] Peter Janich[13] oder Jürgen Mittelstraß[14]), wenn sie von der Notwendigkeit einer Verankerung der Wissenschaftssprache in der Sprachpraxis des Alltags und der Rückbindung der Wissenschaften an die Normen eines vernunftgeleiteten, allen Menschen zugänglichen Prozesses einer öffentlichen Argumentation sprechen. Sie ziehen mit diesem Postulat übrigens aus der Einsicht Ludwig Wittgensteins die passenden Konsequenzen, der seinen frühen Ansatz zur Theorie der „Welt" und den Wissenschaften im „Tractatus" mit seiner Erkenntnis zur Rolle der Sprache in der späteren Schrift, in seinen „Wissenschaftlichen Untersuchungen"[15] deutlich erweitert hatte.

---

10 Vgl. Jürgen Habermas, Theorie des kommunikativen Handelns, 2 Bände, Frankfurt 1981 sowie ders., Vorstudien und Ergänzungen zur Theorie des kommunikativen Handelns, Frankfurt 1984.
11 Vgl. Karl-Otto Apel, Transformation der Philosophie, 2 Bände, Frankfurt 1973.
12 Vgl. Friedrich Kambartel, Jürgen Mittelstraß (Hgg.), Zum normativen Fundament der Wissenschaft, Frankfurt 1973.
13 Vgl. Peter Janich, Zweck und Methode der Physik aus philosophischer Sicht, Konstanz 1973 sowie ders., Friedrich Kambartel, Jürgen Mittelstraß (Hgg.), Wissenschaftstheorie als Wissenschaftskritik, Frankfurt 1974.
14 Vgl. Jürgen Mittelstraß, Das praktische Fundament der Wissenschaft und die Aufgabe der Philosophie, Konstanz 1972 sowie ders., Die Möglichkeit von Wissenschaft, Frankfurt 1974. Vgl. auch den Beitrag von Mittelstraß in dieser Publikation.
15 Ludwig Wittgenstein, Wissenschaftliche Untersuchungen, Frankfurt 2001.

So zeigt sich die klassische Rede der philosophischen Erkenntnistheorie von einem „Objekt" und einem korrelativen „Subjekt" der Erkenntnis und – wissenschaftstheoretisch gesprochen – von „der Welt der Tatsachen" einerseits und den „Wissenschaften" und ihren Forschungsprozessen andererseits als die Bestimmung von aufeinander wechselweise bezogenen und in diesem Sinn „relativen" Größen. „Objekt" und „Subjekt" der Erkenntnis, die „äußere Natur" der Welt im Ganzen als „Gesamtheit aller Tatsachen" und die pluralen Verfahren der Wissenschaften sind aufeinander bezogen in dem Sinne, dass wir nicht von dem einen Moment sprechen können, ohne das andere bereits mitzudenken. Dieser Zusammenhang ist für jedes Erkennen und für jede Wissenschaft konstitutiv, kognitiv unvermeidbar und in diesem Sinn auch notwendig und wissenschaftlich „unhintergehbar". Die Einsicht in die wissenschaftspragmatisch verstandene „Unhintergehbarkeit" der Subjekt-Objekt-Relation und der inneren Beziehung unserer Rede von „Natur" und „Wissenschaft" ist auch ein Hinweis darauf, dass alle Theorien scheitern müssen, die diese Beziehung zugunsten einer der beiden Seiten aufzulösen versuchen. Das gilt für einen „szientistischen Objektivismus", der – wie im Fall einiger prominent gewordener naturalistischen Ontologien in der Gegenwart[16] – die subjektive oder forschungslogische Vermittlung im Blick auf das Konzept von „Natur" unkritisch übersieht. Dasselbe gilt, wenn auch mit umgekehrten Vorzeichen, für den extremen „Subjektivismus", der die Einsicht in die subjekt- und wissenschaftsvermittelte Erkenntnis „der Welt" zum Anlass nimmt, die Möglichkeit von „Wahrheit" oder „objektiver Gewissheit" in den Wissenschaften prinzipiell oder grundsätzlich zu leugnen. An deren Stelle wird dann – vermeintlich kritisch – die Rede von einer „Herrschaft der Technik" (Martin Heidegger[17]) oder von „Machtdiskursen" (Michel Foucault[18]) gesetzt, die dieser Auffassung zufolge den wesentlichen Kern der „Rationalität" der Wissenschaften und der Philosophie ausmachen. Es zeigt sich aber, dass der Anspruch auf „Wahrheit", den die Wissenschaften für ihre Resultate erheben, seinerseits grundsätzlich und wissenschaftsimmanent begründet stets auch auf die bleibende Möglichkeit des „Irrtums" verweist. Aus dieser erkenntniskritischen Einstellung und kognitiven Distanz der Wissenschaften gegenüber ihren eigenen Resultaten folgt aber nicht die Haltung einer Resignation und eine Flucht ins Irrationale, sondern die normativ relevante Einsicht in die Notwendigkeit, alle Erkenntnisse gegenüber berechtigtem Einspruch und methodischem Zweifel immer wieder neu zu rechtfertigen – und in diesem Sinn beständig unter einem Begründungszwang zu operieren. An dieser Stelle aber sehen wir, wie eine kritisch aufgeklärte Rede von „der

---

16 Vgl. hierzu u. a. Wilfrid Sellars, Science, Perception and Reality, Altascadero/Cal. 1963 oder Paul M. Churchland, Scientific Realism and the Plasticity of Mind, Cambridge 1979.

17 Vgl. Martin Heidegger, Die Frage nach der Technik [1949], in: ders., Vorträge und Aufsätze, Frankfurt 1954, S. 13–44 sowie ders., Was heißt Denken?, Frankfurt 1971.

18 Vgl. Michel Foucault, Die Archäologie des Wissens, Frankfurt 1973 sowie ders., Der Wille zum Wissen, Frankfurt 1977.

Natur" nicht nur mit dem Erfordernis eines kognitiven Eingriffs der Wissenschaften „in die Natur", sondern auch mit der Frage nach derjenigen „Normativität" verbunden ist, die dem Erkenntnisprozess der Wissenschaften insgesamt zugrunde liegt. Der Erkenntnisprozess der Wissenschaften ist zentral ein Prozess der permanenten (Selbst)Kritik der Wissenschaften, der auf Dauer gestellt werden muss, aber kein Verfahren, das dazu einlädt, letzte Gewissheiten im Sinne „umfassender Lehren" (John Rawls spricht mit Blick darauf von „comprehensive doctrins"[19]) zu verbreiten. Dies ist nicht die Aufgabe der Wissenschaften, auch nicht die Aufgabe der Philosophie, die auch nur eine fallible Wissenschaft unter den anderen ist, auch wenn sie über bestimmte Prämissen der Wissenschaften ihrerseits nachdenkt und hierzu Vorschläge macht.

In den Wissenschaften vom menschlichen Handeln begegnen wir einem anderen begrifflichen Umgang mit der Frage nach „Natur" und „Normativität" und ihrem inneren Zusammenhang. Dies ist verständlich vor dem Hintergrund, dass wir im Bereich des menschlichen Handelns auf die erweiterte Bedeutung des „Normativen" stoßen, von der bereits zu Beginn meines Beitrag (vgl. 1.) kurz die Rede war, nämlich die Betrachtung der Frage der „Normativität" nicht allein aus der Perspektive einer empirischen Beschreibung und deskriptiv ansetzenden Erklärung von Ereignissen, sondern aus der Perspektive von uns Menschen, verstanden als Handlungssubjekte, die von der Bedeutung von „Normen" als evaluativen und präskriptiven Aussagen unmittelbar (oder auch nur mittelbar) angesprochen und betroffen sind. Diese Erfahrung wird in den Rechtswissenschaften und der politischen Theorie, in der Moralphilosophie und der Moraltheologie reflexiv verarbeitet – und so ist es nicht überraschend, dass wir in der Geschichte gerade dieser Disziplinen auf das Konzept eines „Naturrechts" treffen, in dem – betrachten wir zunächst nur einmal den Terminus – der Begriff der „Natur" und der Anspruch des „Normativen" unmittelbar miteinander verschränkt erscheinen. In der Geschichte dieser Disziplinen wurde mit dem Begriff eines „Naturrechts" oder auch eines „natürlichen Gesetzes" die Vorstellung artikuliert, dass es für das Handeln von uns Menschen sowohl evaluative als auch präskriptive Regeln – und das heißt „Normen" – gibt, die aus bestimmten Gründen gewisse grundlegende Prinzipien etwa für eine „gerechte" politische, gesellschaftliche oder soziale Ordnung einfordern. Dabei kommt dem in diesem Zusammenhang verwendeten Begriff von „Natur" oder auch von „natürlicher Einsicht" ein vielfältiger Sinn zu, der gewisse Ähnlichkeiten zu der bereits dargestellten Vielfalt der Bedeutungen von „Natur" aufweist.

Begriffssystematisch ist es in hohem Maß relevant, von welcher Bedeutung im Begriff der „Natur" die von der Rechtslehre, von der politischen Theorie oder von den Ethiken in Anschlag gebrachte Rede des „Naturrechts" ausgeht. Versteht man nämlich unter „Natur" nur die Angabe oder Definition einer „wesentlichen Eigenschaft", wie wir dies als eine erste Bedeutung der Rede von „Natur" festgehalten hatten, dann be-

---

19 Vgl. John Rawls, Politischer Liberalismus, Frankfurt 1998.

zeichnet der Begriff des „Naturrechts" auch nur diejenigen Eigenschaften, durch die die Idee des „Rechts" ihrerseits bereits selbst „wesentlich" bestimmt ist. Jede Rechtsordnung ist dann „natürlich" und entspricht dem angenommenen Begriff des „Rechts", die in einer wie auch immer gearteten gesellschaftlichen Ordnung eine regulatorische Ordnungsfunktion effektiv erfüllt, ganz gleich, welche Ordnungsziele oder weiteren Präferenzen von dieser Ordnung favorisiert werden. Versteht man den Begriff des „natürlichen Rechts" in diesem Sinn, dann lässt sich jedwedes auf Rechtsregeln basierte System als eine „Rechtsordnung" positiv auszeichnen. Die Deutung des Rechts als eines „auto-poietischen Systems" mit funktionaler Steuerungsfunktion bei Niklas Luhmann[20] weist durchaus eine gewisse Nähe zu dieser Fassung des Begriffs des „natürlichen Rechts" im Sinn seiner wesentlichen Merkmalsbestimmung auf, auch wenn Luhmanns Systemtheorie des Rechts ihrerseits nicht auf den Begriff eines „natürlichen Rechts" zurückgreift.

Versteht man aber dagegen unter dem Begriff der „Natur", wie wir gesehen hatten (vgl. 2.), einen Ausdruck für eine Sphäre des „Vorgegebenen", des „Ursprünglichen" und vor allem des „Allgemeinen" und somit für das, was „überall" vorhanden und gleich ist und was sogar „dieselbe Kraft hat"(Aristoteles),[21] dann beschreibt das Wort „Naturrecht" im Kern nichts anderes als diejenigen „Rechte", die zumindest „tatsächlich" allgemein sind und überall auf der Erde gelten. Aristoteles führt in seiner Nikomachischen Ethik in der Tat den Begriff des „natürlichen Rechts" analog zu den Gesetzen der Natur ein, die überall auf der Erde gleich sind, genauso wie „das Feuer hier und in Persien brennt". Dieses Verständnis hat für lange Zeit den Begriff des „Naturrechts" geprägt, so insbesondere in der Geschichte des „Völkerrechts" (lat.: „ius gentium"). Hier wurden alle diejenigen Rechtsregeln, die überall auf der Welt gelten, zum Bestandteil des „Völkergewohnheitsrechts" und zu einer „Rechtsquelle" erklärt. Von dem überall faktisch geltenden Recht wurde in der Theorie des Völkerrechts dann lange Zeit behauptet, es sei im Kern genau das, was das „Naturrecht" oder das „natürlich Rechte" ausmache. So konnte man in der Rechtsgeschichte und insbesondere im Anschluss an die Rechtslehre bei Cicero, der in seiner Schrift „De legibus" von der „vera lex"[22] spricht, in der bis weit ins Mittelalter und die Neuzeit fortgeführten Tradition des römischen Rechts mitunter selbst noch das Verhältnis der Tiere zueinander unter den Begriff eines „natürlichen Rechts" subsumieren.[23] Hier können wir beobachten, wie Motive der stoischen Naturphilosophie mit allgemein bekannten Rechtsregeln amalgamiert wurden und zur Konzeption eines „Naturrechts" führten, die für uns

---

20 Vgl. hierzu Niklas Luhmann, Das Recht der Gesellschaft, Frankfurt 1975.
21 Aristoteles, Nikomachische Ethik, Buch 5, Kap. 10, 1134 b 25.
22 Vgl. Marcus Tullius Cicero, De legibus I, 18.
23 Vgl. hierzu die Aufnahme der Definitionen des Naturrechts bei den römischen Rechtsgelehrten Gaius und Ulpian in den Digesten, I, 1 sowie I, 2.; vgl. hierzu auch Isidor von Sevilla, Etymologiarum sive orgini librorum XX, Buch 5.

heutige Rezipienten der Geschichte des Naturrechts äußerst merkwürdig erscheint. Das ändert nichts an der hohen Wirksamkeit dieser Rechtskonstruktion. Sie ist historisch z. T. dadurch zu erklären, dass Augustinus das stoisch inspirierte Naturrecht bei Cicero mit der christlichen Prämisse rezipierte,[24] der zufolge das „natürliche Recht" ein Schöpfungswerk Gottes sei, das allen menschlichen Rechts- und Moralordnungen normativ vorgeordnet war. Diese Verbindung von stoischer Kosmologie, römischer Rechtsgelehrsamkeit und christlicher Theologie gab der Tradition des „Naturrechts", in der „die Natur" mit dem Willen des Schöpfers identifiziert wurde, eine lange Wirkungsgeschichte. Doch es waren gerade die Philosophen, Theologen und Juristen an den Universitäten des Hoch- und Spätmittelalters, die diese Tradition des „klassischen Naturrechts" argumentativ aufbrachen und grundlegend veränderten, so dass sich aus dem „Naturrecht" die Tradition eines „Vernunftrechts" entwickelte, in dem allen Menschen das „natürliche Vermögen" der praktischen Vernunft zugesprochen wurde, auf Grund eigener Einsicht das moralisch Gute und das politisch Gerechte zu erkennen.

Den beiden bisher hier genannten Lesarten des Begriffs eines „natürlich Rechten", also der funktionalen Definition der „Natur" oder des Wesens des „Rechts" sowie der Auslegung des Naturrechts als der allerorten verbreiteten *faktischen* Geltung von gewissen „Rechtsprinzipien", fehlt eine klare Vorstellung davon, dass dem Begriff des „Rechts" selbst ein *normativer* Anspruch auf Geltung zukommt. Damit ist gemeint, dass es „im Recht" selbst ein Anspruch auf Geltung im Sinn einer vernünftig begründeten Normativität angelegt ist, das das „gerechte Recht" von jedwedem Willkürrecht in der Hand des Stärkeren grundlegend unterscheidet. Beide bisher genannten Vorstellungen begnügen sich damit, das Recht auf seine Ordnungsfunktion zu reduzieren und fragen nicht danach, ob es zwischen einer „ungerechten Rechtsordnung" und einer „gerechten Rechtsordnung" einen Unterschied gibt, der im Recht selbst begründet werden kann. Dieser Gedanke setzt einen neuen, bei Cicero erst schwach begründeten Gedanken voraus, den wir als das „vernunftnormative Argument" im Recht bezeichnen können. Dieses Argument wird im Verlauf einer langen Debatte über Sinn und Bedeutung des Konzepts eines „Naturrechts" allmählich entwickelt. Es basiert auf der Annahme, dass das Recht eine Aufgabe wahrnimmt, die gleichsam in der praktischen Vernunft selbst wurzelt, einer Vernunft, die prinzipiell allen Menschen zugänglich ist. Die praktische Vernunft, die sich dieser Auffassung zufolge in den Forderungen des „Naturrechts" artikuliert, hat die Funktion, das geltende Recht auf bestimmte Regeln festzulegen und jede Form von rechtlicher Willkür zu beschränken. Genau diese Argumentation bestimmt die elaborierte Konzeption eines „Naturrechts" bei Thomas von Aquin, wobei im Blick auf ihn zwischen einem Begriff des „Naturrechts" (lat.: „ius

---

24 Vgl. hierzu u. a. Aurelius Augustinus, De ordine, 2, 8 sowie De libero arbitrio, 1, 5–7 und De vera religione, 30.

naturale") und einem Begriff des „natürlichen Gesetzes" (lat.: „lex naturalis") unterschieden werden muss.[25]

Unabhängig von weiteren Differenzierungen im Detail können wir feststellen, dass bei den vernunfttheoretisch argumentierenden Rechtstheorien seit dem hohen Mittelalter von einem juristisch, ethisch wie politisch relevanten „Naturrecht" auf der Basis der vernünftigen Einsicht von uns Menschen die Rede ist. Das so verstandene und von den zuvor genannten Bestimmungen verschiedene „Naturrecht" artikuliert die Forderung, dass vernünftige Maßstäbe an das System des Rechts, den Bereich der Moral und den Raum der Politik anlegt werden müssen. Das „Naturrecht" bezeichnet, in diesem Verständnis, eine Reihe von Prinzipien, die für jedes „legitime Recht" gelten sollen, wobei die Idee der „Legitimität des Rechts" aus der praktischen Vernunft abgeleitet wird, mittels derer das faktische, also das rein „legale" oder „positive" Recht kritisiert und verändert, auch geschichtlich weiterentwickelt werden kann. Das durch die Vernunft begründete „Naturrecht" weist somit bestimmte Einsichten als Bestandteile eines gleichsam „vor-positiven" Rechts aus, womit zugleich der Unterschied zwischen dem in einem politischen Gemeinwesen geltenden „positiven Recht" und den vom „Naturrecht" reklamierten, normativ gültigen und in diesem Sinn „vor-positiven" Rechtsprinzipien eingeführt wird, ohne den nicht einmal der große Theoretiker des modernen Rechtspositivismus Hans Kelsen auskommt. In dieser Differenzierung wurzelt die für das Recht als Instrument der Regelung des äußeren Handelns von uns Menschen bis heute wichtige Unterscheidung der Legalität und der Legitimität des Rechts.

Mit der Annahme einer auch rechtlich verbindlichen Geltung von sog. „ersten" oder „obersten Rechtsprinzipien" im klassischen Naturrecht ist auch die Vorstellung verbunden, dass die ins Einzelne gehende Regelungskraft des „positiven" und in diesem Sinn von einem Gesetzgeber „gesetzten" Rechts entweder aus den obersten Rechtsprinzipien logisch abgeleitet werden kann, oder aber, dass das „positive Einzelrecht" der Staaten den obersten Prinzipen des Rechts zumindest nicht diametral widersprechen darf. Eine solche normative Lesart des Begriffs eines „vor-positiven" Rechts wurde in der späteren Rechtsgeschichte der Aufklärung auch als „Vernunftrecht" bezeichnet, weil und insofern man davon ausging, dass die allen Menschen „natürlicherweise" zugängliche, also die allgemeine Menschenvernunft im Erkennen und Handeln eine Einsicht in die ersten Prinzipien dessen eröffnet, was als gerecht und als ungerecht gelten kann. Hier lässt sich ablesen, dass mit Blick auf das Denken des Rechts insgesamt von einer „Grundnorm" (Hans Kelsen)[26] gesprochen werden kann, auch wenn wie im Fall des modernen Rechtspositivismus bei Hans Kelsen keine vor- oder überpositiven Rechtsbestände in die geltende Rechtsordnung einwandern sollen. Doch selbst für

---

25 Vgl. hierzu u. a. Thomas von Aquin, Summa theologica I–II, qq. 90–95.
26 Vgl. Hans Kelsen, Reine Rechtslehre [1934], Tübingen 2008, S. 73 ff. und S. 93.

den Rechtspositivismus gilt die Einsicht, dass eine „Grundnorm" anzunehmen ist, die allen weiteren Rechtsnormen oder juristischen Einzelgesetzen normativ vorausliegt.

Die Frage der Möglichkeit oder Unmöglichkeit der Annahme eines Natur- oder Vernunftrechts hat in den modernen Debatten der Rechtswissenschaften und der Rechtsphilosophie zu vielen Kontroversen, aber auch zu manchen Missverständnissen geführt, die uns hier nicht im Einzelnen beschäftigen können. Im Kern enthält die Annahme, dass ein Natur- bzw. Vernunftrecht möglich ist, die These, dass die Rede von einer „normativen Geltung" des Rechts unmittelbar mit einem Rekurs auf ein „erstes Prinzip" allen Rechts oder auf oberste Rechtsprinzipien wie im Fall der Tradition des klassischen „Naturrechts" verknüpft ist. In keinem Fall aber liegt im Blick auf diese These ein sog. „naturalistischer Fehlschluss" vor. Ein „naturalistischer Fehlschluss" liegt genau dann vor, wenn man versucht, aus etwas „Vorhandenem", das wir deskriptiv erfassen können, zugleich eine allgemeine präskriptive Regel oder eine Norm für das menschliche Handeln abzuleiten. Eine solche Argumentation aber liegt sowohl dem Natur- wie dem Vernunftrecht fern; denn in beiden Fällen referieren die Begriffe „Natur" bzw. „Vernunft" nicht auf etwas, was in einem schlichten Sinn „vorhanden", faktisch „gegeben" oder auch autoritativ „vorgegeben" ist. Das Natur- oder Vernunftrecht im hier erläuterten Sinn ist vielmehr gerade dadurch definiert, dass die Einsicht in etwas „Gesolltes" erst aus der „Vernunft", der „vernünftigen Einsicht" der Beteiligten erwächst und nicht aus vermeintlichen oder wirklichen „Fakten" einfach abgeleitet wird. Dem „Gesollten" kommt somit, da es aus der vernünftigen Argumentation hervorgeht, der Charakter eines in gewisser Weise „Kontra-Faktischen" zu. Es erwächst aus der Einsicht einer Person, die das „Normative", das „Gesollte" aus der Betrachtungsperspektive der „ersten Person" („Ich soll … / Wir sollen …") erkennt und als etwas für die eigene Vernunft, den eigenen Willen und das eigene Handeln Verbindliches identifiziert. Das macht deutlich, dass bereits die Rede von der „Natur" im klassischen Naturrecht nur als eine Referenz auf eine Gestalt von „vernünftiger Normativität" qualifiziert werden kann.

Aus den Handlungswissenschaften, insbesondere der Rechtswissenschaft, aber auch der Poltischen Theorie, der Moralphilosophie oder der Moraltheologie, aus Pädagogik, Psychologie und Medizin ist zwar nicht der Begriff oder der Terminus, sehr wohl aber das sachliche Gebot eines „aus Gründen der Vernunft" Richtigen – und das heißt eines „von Vernunft aus normativ Gesollten" – nicht zu eliminieren. Das sagen uns heute zumindest Vertreter und Vertreterinnen der Philosophie in der Nachfolge von Immanuel Kant. Für Kant gehört das Postulat der „Freiheit" eines jeden Menschen, verstanden als seine sittlich-moralische und seine politische Autonomie, zum Bestand des „unbedingt Gebotenen". Konsequenterweise bezeichnet Kant diese Forderung in seiner späten Rechtslehre mit dem Begriff eines „Menschenrechts" auf Freiheit.[27] Analoges

---

27 Immanuel Kant, Metaphysik der Sitten, Rechtslehre, Akad.-Ausgabe, Band VI, Berlin 1968, S. 237.

bringt der Begriff der „Würde" eines jeden einzelnen Menschen zum Ausdruck. Auch in der Gegenwart lässt sich die Forderung einer „unbedingten Geltung" der allgemeinen Menschenrechte, zumindest aber der formal negativ gefassten allgemeinen Verbotsrechte aus dem Katalog der allgemeinen Menschenrechte, nicht ohne den Gehalt der Bedeutung einer Rede von einem „natürlichen Recht" eines jeden einzelnen Menschen auf Unterlassung von erheblichen Einschränkungen seiner Freiheit explizieren. Dabei liegt die Pointe der Aussage gerade darin, dass sich das intendierte „Normative", nämlich die unbedingte, universale und somit praktisch nicht eingeschränkte Geltung eines Verbots/Gebots, nicht anders als durch den Hinweis auf die „Natur" des legitimen, des „gerechten Rechts" und die besondere Verfassung der Menschen artikulieren lässt. Diese liegt wiederum in deren „Natur" als vernunft- und sprachbegabter, als freier und gleicher Wesen begründet. Aus dieser Einsicht hatte bereits Kant[28] die Forderung abgeleitet, dass einem jeden Menschen, gleich welchen Alters, gleich welchen Geschlechts, gleich welcher sozialen und gesellschaftlichen Stellung sowie kulturellen und ethnischen Herkunft eine singuläre, eine unaufgebbare Würde zukommt, die zwar faktisch mit Füßen getreten werden kann, deren normatives Erfordernis aber von niemandem außer Kraft gesetzt werden kann.

Matthias Lutz-Bachmann
Johanna Quandt Young Academy / Institut für Philosophie der Goethe-Universität

---

28 Vgl. Immanuel Kant, Grundlegung zur Metaphysik der Sitten, Akad.-Ausgabe Band IV, Berlin 1968, S. 426–440; vgl. hierzu u. a. Matthias Lutz-Bachmann, Art. „Würde", in: Staatslexikon, 8., völlig neu bearbeitete Auflage, Bd. 6, Freiburg 2021, Sp. 472–478.

About the JQ Young Academy

ELENA WIEDERHOLD

Different levels of participation, together with the mixture of different disciplines and universities, ensure the JQYA's international and interdisciplinary character and create a long-lasting international network of researchers.

> Goethe University wants to tap the potential of young scientists (…), recognise the most talented and dedicated scientists and train them to become global scientific leaders.
>
> *Prof. Dr Enrico Schleiff, President of Goethe University and Founding Director of the JQYA*

Recognising that young scientists are the driving force in the formulation of novel ideas and concepts for all aspects of the university – research, teaching, structures, and visions for the future – Goethe University founded the interdisciplinary Johanna Quandt Young Academy (JQYA). The establishment of the JQYA was made possible by the generous donation from the Quandt family to the University Foundation. With the Johanna Quandt Young Academy opening in 2018, Goethe University launched an innovative structure for the promotion of early career researchers – postdocs, assistant professors, group leaders – at the beginning of their scientific independence and during their professional qualification. This transition phase is a, if not the, crucial time in career development to a professorship and a role as a leading scientist. There are several Young Academies in Germany and in Europe, but not within a university structure. Thus, the Johanna Quandt Young Academy is a visionary concept ahead of its time because it integrates an academy as a supportive and educative element within the university's infrastructure. Goethe University is, therefore, the only German university that, through the platform of the Academy together with cooperation partners, offers versatile support based on different interlocking mechanisms.

Core of the Academy Programme: Academy Theme

The JQYA displays the ideals of traditional academies by being a learned society providing a platform for intellectual exchange across disciplines. To integrate the different levels of participation and different disciplines present in the Academy's Programme, an annual Academy Theme based on a global topic has been established. A global topic is defined by the Academy's members in a workshop each year such that all participating disciplines and generations can contribute to discussions equally. A young researcher's career development benefits from their integration into the Academy Programme because it allows them to learn about other disciplines' approaches, increase awareness of their own research within a broader context and, therefore, gain a better understanding of their own research questions through scientific dialogues and cross-disciplinary debates. All this inspires creative solutions and extends scientific horizons.

Integrative nature of JQYA: different participation levels

The JQYA offers equal access to academics from a wide range of backgrounds, disciplines, nationalities, generations and genders. From the 24 fellows and members, every third person originates from a different country, nearly 40 % are female scientists, and every second scientist comes from a different discipline. With English as the working language, the JQYA connects young scientists with outstanding leaders from the international scientific community from the main fields of life sciences, humanities and social/political sciences, as well as arts and literature. Every three years, the founding directors and the Executive Board of Goethe University appoint up to four distinguished senior scientists as the external board of the JQYA. These renowned scientists are established international researchers with extraordinary scientific achievements and unprecedented commitment to young scientists. They play a crucial role by giving support and advice to the Academy's directors and by delivering critical and inspiring impetus to the fellows' scientific discussions.

Young researchers can join the JQYA each year as fellows/international fellows or as members. Advanced postdocs, habilitating scientists, independent group leaders, assistant and qualification professors without a permanent contract at Goethe University and partner institutions can apply for fellowship status. Up to 10 of the most excellent candidates from different disciplines and fields are selected by the distinguished senior scientists and confirmed by the JQYA Board of Directors each year to become **JQYA fellows**. Fellows are integrated into the platform of the Academy Programme and shape the inner core of the Academy, drive cross-disciplinary discussions within the Academy Programme, and actively contribute to the organisation's goals and activities. Furthermore, fellows direct the academic work by annually formulating the

Academy Theme (e. g., 'Nature and Normativity', see below), which aims at intercon-
necting and strengthening cross-disciplinary work at a national and international level.
In the first three rounds, the JQYA gained 19 outstanding early career researchers from
11 disciplines affiliated with 14 departments.

JQYA members form the next layer of JQYA participants. Membership is a part
of Goethe University's strategy to acquire the best internationally proven early career researchers from abroad or from other German universities and affiliate them to
Goethe University in the important transition phase from a group leader to a call for
a professorship. Internationally renowned junior group leaders and junior principal
investigators who have successfully applied for highly competitive third-party funding, such as the Emmy Noether Programme or the European Research Council Starting Grant, and are in the process of establishing their own research group at Goethe
University, can apply for JQYA membership. Membership duration is linked to the
candidate's third-party funding duration and is applicable for at least five years. Only a
few successful applicants each year are granted this solid financial support through the
JQYA. Over the past three years, we have attracted five members to Goethe University.
Although members are not obliged to participate in the Academy's programme, we are
proud to note that all members actively participate in the Academy's events with great
enthusiasm.

Fellows and members will become **JQYA alumni** upon expiry of their member-
ships. In the section 'Participants', we present our JQYA fellows, JQYA members and
distinguished senior scientists in more detail.

Senior scientists at Goethe University can join the JQYA as **'ambassadors'** upon
invitation from the JQYA directors, facilitating the networking and integration of the
younger generation into Goethe University's community.

Needs-oriented support lines

Besides being a platform for the exchange of scientific ideas, JQYA offers significant
financial support. Individually tailored support lines guarantee scientific freedom to
pursue an academic career more efficiently. There are four support lines offered by
the Academy Programme: **(I) Scientific Development** which provides access to the
educational elements of the Academy, offered in agreement with the fellow's needs;
(II) Academy Fellowship is awarded for the purpose of implementing research activities, along with carrying out one's own research project as per the Academy Theme;
(III) Sabbatical Fellowship releases the fellows from their teaching duties for one
semester, allowing them to focus on their own research activities; and **(IV) Internationalisation Stipends** (former: International Academy Fellowship) that is meant for
establishing international collaborations, allowing fellows a short research stay abroad
or inviting guest researchers to Goethe University.

An additional line for members, **Science Funding**, is available to incoming group leaders with their own self-funded positions. The Science Funding is to support members in the further development of establishing an independent research group. The funding is dedicated to expenses that other funds cannot finance, ensuring the candidate has the best possible financial flexibility and freedom.

With opportunity comes responsibility

In 2021, to incorporate the Academy into the university's infrastructure, the Presidential Board of Goethe University transferred the duties of responsibility of the Early Career Researcher's Council into the hands of the JQYA's fellows and members. By doing this, the President's Office acknowledges the JQYA's commitment to Goethe University and allows it to act as an advisory body to the president. In future, early career researchers will co-shape the university and be more closely involved in the university's politics. On the other hand, the President's Office can respond directly by designing the university's education policies in line with young researchers' current and future needs.

Academic work of the JQYA

ELENA WIEDERHOLD

The Academy Programme and the Academy Theme provide the foundation for the JQ Young Academy's academic work. With 14 disciplines currently represented in the JQYA, a global Academy Theme is set by the JQYA members in such a way that it can be reflected by any of these disciplines.

The joint thematic development of the Academy Theme in a cross-disciplinary manner is tackled within versatile elements of the Academy Programme. In this section, we report on different formats and present the most relevant pieces of the Academy's work to the respective Academy Theme.

Platforms of the JQYA's academic work

- **JQ Academy Day** offers academics at Goethe University an opportunity to gain insights into the JQYA's interdisciplinary work towards a common theme and encourage more profound and creative thinking beyond the *wet lab*.
- **The Award Ceremony** is an annual event to welcome and award new members.
- **The Summer School** completes the academic programme and usually closes an academic year. Summer schools aim for an intense exchange between the members and fellows of the Academy and invited guests.

Formats of the JQYA's academic work

The work of the Academy has different formats, and the programme is not repetitive but rather can change every year along with the annual theme. The driving and creative force behind the dynamic changes are the fellows and members. In the past three years the JQYA has exploited and tested the following formats:

–   **Roundtable discussions consisting of:**
'**Comprehending the interdisciplinarity**' session aims at presenting one's own research in a comprehensive manner and comparing the research approaches with other disciplines.
'**Ask me anything**' is an interactive session to encourage cross-disciplinary basic questions that will help to understand other research disciplines.

–   '**Getting to know each other**' is an introductory event aimed at facilitating the first contacts among members, which enables the forming of scientific tandems and cooperation within the Academy`s programme.

–   **Workshop** '*Academy Theme*' is an annual event that aims at the developing and defining of an annual topic.

–   **Study days** are devoted to the internal discussion of the global topic and the production of the joint project. The result of Study Days 2020–2021 is a joint paper that was optimised for an oral keynote speech on 'Tenets, Variations, Transformations' and a written essay in German 'Grundätze, Variationen, Transformationen. Überlegungen zur Arbeit und Rolle der Johanna Quandt Young Academy in Wissenschaft und Gesellschaft'.

–   **Scientific tandems,** in which two to three members join together to organise scientific events. Such tandems have resulted in the Tandem-Symposium in Brussels, Panels at the Academy Day and the public discourse 'Frauen in der Politischen Öffentlichkeit'.

–   **Presentations, seminars, lectures and book presentations,** which regularly take place on the JQYA platform.

–   **Joint events** in cooperation with other partners of Goethe University: the JQYA annually participates in the Postdoc Day of the Rhein-Main-Universities' Alliance.

–   **Director's invitation** is a series of lectures with a concluding discussion round. Invited speakers are established senior scientists at Goethe and partner universities, so-called ambassadors.

II.   2018–2019 Nature and Normativity

The academic work of the JQYA started in 2018 with the Academy Theme dedicated to 'Nature and Normativity'. In the history of the sciences and humanities, we can identify a mutual interference and understanding of the concept of nature and the norms for human acting. Under the conditions of modern sciences, JQYA's fellows reflected on the understanding of 'nature' or of the 'natural' from different perspectives. The natural scientists realised that their understanding of 'nature' differs from that of a philosopher.

–   What do we mean when we define a regularity in nature as 'Natural Law'?
–   How is the understanding of 'nature' or of 'natural' constitutive for our reading of 'normativity'?
–   What are the relations between the supposition of 'nature' and the concept of 'normativity' in the different fields of our disciplines?

These and similar questions defined the focus of the interdisciplinary discourse with which we opened the scientific programme and research studies of the JQYA.

Guest Keynote Lecture 'Nature and Normativity'

Jürgen Mittelstraß

At the Opening Ceremony on September 19, 2018

Nature and normativity belong to the key concepts of philosophy, particularly in the form of conceptual opposites like nature and mind, nature and history, nature and culture – to the most difficult ones. In the case of normativity, being the source and the measure of all ethical systems, this implies, anyhow, in the case of nature, the epitome of life, as well. The same is true of the connection between both, such as in the correspondence between *nomos* (law) and nature in classical Greek philosophy. Even after the books of philosophy of nature were closed and modern science started its triumph march, both concepts, nature and normativity, remained affiliated with each other, albeit in a superficial way – such as in the legal field in the concept of natural law (*Naturrecht*, which, within the framework of a philosophical reconstruction, means the concept of rational law: *Vernunftrecht*). Against the background of science, nature loses its character in orienting life and thereby ethicists, one of its previous foundations. The question is whether this needs to remain or rather whether nature, in some other way, may still provide guidance in an ethical sense.

Concerning ethics, ethics is always normative in so far as it has neither a purely empirical foundation nor can it be organized solely in the form of meta-ethics. Making ethical judgments on an empirical basis leads to ethical naturalism,[1] the effort to operate ethics only as meta-ethics, i. e. in a linguistic analysis of ethical judgments, into ethics without morality. Here, as well, the question to what extent recourse to nature, i. e., to natural conditions, may serve as an authority for ethical conceptions, whether of a descriptive or prescriptive (normative) kind, remains open. In the following, a few systematical remarks on this subject could be described as leading from an anthropo-

---

1 On the concept of ethical naturalism see R. Wimmer, 'Naturalismus (ethisch),' in: J. Mittelstraß (Ed.), Enzyklopaedie Philosophie und Wissenschaftstheorie, 2nd edition, vol. V, Stuttgart and Weimar: Verlag J. B. Metzler 2013, pp. 504–506.

centric ethical concept to an ecological vision and back. But first of all, to the keyword 'nature'.[2]

1.   From a natural paradigm to a technological paradigm

Modern societies have problems with nature due to the rupture between nature and (rational) life typical of these cultures. Rational man no longer lives with or within nature; he faces it in his modes of work and production. He does not belong to nature: nature belongs to him. Culture has made this change the mode of living typical of man and the reality of nature. Nature itself becomes part of a rationally organized reality; it acquires a new quality. The result of the development of natural science and technology in the modern age is not just a new society, the 'technological' society or the 'technological' culture, but also a new nature, the 'technologically appropriated' nature. Francis Bacon's slogan that one can take possession of nature by following its laws,[3] and Hegel's proposition that man has 'in his tools (…) the power over external nature,'[4] are realized in technological culture.

This development also involves replacing the *nature paradigm* for all orders, including the order of human life, with a *technological paradigm*. This means that today, discoveries in the natural sciences are primarily due to the conditions defining a technological practice and no longer representing a natural order. Thus, the modern natural sciences, as laboratory sciences, mainly produce their objects to examine them either after or during their production. Decisive in our context is that modern science, especially physics, as opposed to classical (Aristotelian) physics, which sought to construct an *orientational knowledge in nature*, wants to control or utilize nature or provide the scientific preconditions for such a *dispositional knowledge of nature*. To put this differently: in the cosmological or nature paradigm of Greek philosophy of nature, whose object of research was to all appearances an ordered nature, nature and life are inextricably intertwined. Nature remains as it is. In contrast, in the technological paradigm of modern science, life and nature are separated in the orientational sense mentioned before; a new nature comes into existence – without any orientational meaning. A *history of progress* (*Fortschrittsgeschichte*), namely the gradual shift from a cosmological to a technological paradigm in natural research, also reveals in the process elements of a *history of loss* (*Verlustgeschichte*): technological cultures that come into existence in the

---

2 The following is closely based on an earlier account: 'Ethics of Nature,' in: W. R. Shea and B. Sitter (Eds.), Scientists and Their Responsibility, Canton M. A.: Watson Publishing International 1989, pp. 41–57.
3 Novum organum (1620) I 3, The Works of Francis Bacon, vols. I–XIV, ed. J. Spedding et al., London: Longmans and Co. 1857–1874, vol. I, p. 157.
4 G. W. F. Hegel, Wissenschaft der Logik II, Saemtliche Werke. Jubilaeumsausgabe, vols. I–XX, ed. H. Glockner, Stuttgart: Fr. Frommanns Verlag 1927–1930, vol. V, 226.

wake of this transition have enormous difficulties living 'rationally' with nature; given their overabundance of dispositional knowledge (*Verfuegungswissen*) and their relative lack of orientational knowledge (*Orientierungswissen*).

This is, moreover, also the origin of the peculiar problem of being able to precisely say what and where nature is. We celebrate it as a green world that should be before our windows but has long been missing as 'untouched' nature. As a rule, we look at nature merely as a source of raw materials managed by economic and technical expertise, or as the ground in which technological culture puts its waste products, or as a haven for free time that the tourist industry fills with its false dreams. Nature, again, has become a part of technological culture, a part of 'spaceship earth,' in the resolute language of technological culture. Wherever one goes 'in nature,' the knowing, constructing, economizing, and the ravaging mind has already been there. It is the environmental problems that are becoming increasingly stronger and irrefutable, e. g., climate problems, which bear testimony to this development. In his dealing with nature, man confronts himself – increasingly powerful and, at the same time, paradoxically, increasingly helpless.

'Evolutionary' reasons, too, can be given for this view of things – the domination of nature and the loss of nature. While the phylogenetic process of development among living creatures generally occurs as a process of adaptation to their respective environments, this is different for human beings. During his cultural development, man remains virtually unchanged as a biological species and has adapted the environment (and hence nature) to his needs, which means simply to himself. Man, so controls, especially in technological cultures, the arrangement and organization of his environment that it is no wonder that the appearance of limitless independence and perfection, being dominant in these cultures, also influences how he deals with nature. As already emphasized, natural conditions are no longer valuable as orientational elements that can be implemented in life but as that realm where man achieves mastery of his ability to utilize and control. Of course, man's superiority can also be conceived as the ruthless completion and enhancement of his 'nature'. His rationality – understood here (and also a little misunderstood) as a product of evolution – is a part. In human culture, i. e., in man's way of appropriating and changing the world into his world, we find an unmistakable expression of man's nature and ecological success. Even his problems are testimony to this:

> If this magnificent ecological success of our species (…) causes us increasingly more problems and at the same time to all of nature around us, then it is not because we would have strayed from the path of natural virtue, but because up to now we have been following it almost with blind determination.[5]

---

5 H. Markl, Natur als Kulturaufgabe: Ueber die Beziehung des Menschen zur lebendigen Natur, Stuttgart: Deutsche Verlags-Anstalt 1986, p. 358.

In man's 'nature' and his 'earthiness' too, however, something like *the natural* is no longer recognizable – and not only because nature and culture are no longer clearly divisible areas. That nature itself has a history, i. e., a history that is not only part of the history of man, also means that it does not exist as something unalterably natural, as the great being behind all other beings that belong to our world. It was the notion of an *active* and *acting*, sometimes even divine nature, a nature like an intelligent agent, guided by rational and economic points of view, that nourishes and teaches man to survive in his world by imitating its abilities and its purported wisdom. Even Galileo, at the beginning of the modern age, subscribed to this notion at times, for instance when he said 'that God and nature are concerned with the government of humans affairs.'[6] In the context of natural science, like Robert Boyle, a physicist and one of the founders of the *Royal Society*, pointedly phrased it as late as 1686: Nature was considered to be

> a most wise being, that does nothing in vain; does not miss of her ends; does always that, which (of the things she can do) is best to be done; and this she does by the most direct or compendious ways, neither employing any things superfluous, nor being wanting in things necessary; she teaches and inclines every one of her works to preserve itself.[7]

The further development of natural science soon left little of this notion of nature. For Boyle himself, the notion had scarcely more than metaphorical meaning.

This also means that nature comes and goes: nature looks different over long periods of time. Neither the nature of the older palaeolithic era, which provided pre-Neandertal man with scant nourishment nor the younger palaeolithic era, in which *homo sapiens* entered the scene, is *our* nature. Our nature is – and not just within the limits of technological culture – a 'cultural nature',[8] a nature created by man, that is, by his continual interaction with 'natural' facts. All that is 'natural' is a man for himself and the world he has utilized, exploited, cared for, praises, and fears, a world that is never the 'pure' world of nature, but always man's world. In other words, we are mistaken when we oppose the *cultural* to the *natural*, at least when we mean that something *natural* is the *original*, the timeless, the measure of all things, even the human.

---

6 G. Galilei, Dialogo sopra i due massimi sistemi del mondo (1632) III, in: Le Opere di Galileo Galilei. Edizione Nazionale, vols. I–XX, Florence: Barbèra 1890–1909, vol. VII, p. 394.
7 "A Free Enquiry into the Vulgarly Receiv'd Notion of Nature" (London 1686), in: The Works of the Honourable Robert Boyle, 2nd edition, vols. I–VI, ed. Th. Bird, London: J. and F. Rivington et al. 1772, vol. V, p. 174.
8 See H. Markl, Oekonomie und Oekologie: Wissenschaftliche Forschung und oekologische Herausforderungen (Speech given in Mainz on May 13th, 1987 at the 75th Jubilee of the Industrial Association for the Protection of Plants (IPS)), separate publication, p. 6.

2.   Anthropocentric and ecological aspects

Ethical questions in the context of nature and normativity arise when, on the one hand, dealing with responsibilities towards future generations, namely their conditions of life, and, on the other hand, when thinking of responsibilities towards nature itself. Responsibilities towards future generations consist of ensuring that future generations' life spaces or the corresponding options are not restricted irreversibly (keywords: rational economy of resources and preservation of biodiversity, i. e., diversity of species, genetic diversity and diversity of ecosystems). And this is realized by every generation's doing the obvious, that is, passing on the earth to the next generation as they found it, perhaps even a little better and more stable. Here, we speak of an ethics of long-term responsibility.

Another question, and a more difficult one to answer, is whether there is or can be a responsibility towards nature *for its own sake*. With this question, we enter the realm of a *teleological* conception or an anthropomorphic view of nature. Nature appears as a subject with its own aims and rights. The philosopher Robert Spaemann speaks in this connection about the 'moral right of nature' as the 'right of a natural being to be treated in such a way that we can treat it as a good for its own sake.'[9] Hans Jonas states that we 'do not have the right to choose or even to risk the non-existence of future generations because of the existence of the present one.'[10] For Klaus M. Meyer-Abich, also a philosopher, man is related not only to animals and plants, but also to the elements, earth, water, air, and fire: 'In all of nature they are like us, and we are like them.'[11]

These conceptions go too far, at least with respect to the limits of ethical arguments. They do not even apply to man if what we mean here is not only the right of future generations to humane living conditions but, over and above this, a right to exist. If mankind decided recently not to have any more children – naturally, a completely fictitious idea – much could be said against such a decision (from a violation of God's plans for mankind to the eventual loss of social security). We could not say, however, that the non-existent has a right to exist. But what counts for man counts even more for non-human life and for nature as a whole.[12]

---

9 Lecture quoted by H. Lenk, "Verantwortung für die Natur: Gibt es moralische Quasirechte von oder moralische Pflichten gegenüber nichtmenschlichen Naturwesen?," Allgemeine Zeitschrift für Philosophie 8 (1983), p. 3. See also R. Spaemann, Philosophische Essays, Stuttgart: Philipp Reclam jun. 1983, p. 57.
10 Das Prinzip Verantwortung: Versuch einer Ethik für die technische Zivilisation, Frankfurt: Suhrkamp 1979, p. 36.
11 K. M. Meyer-Abich, Wege zum Frieden mit der Natur: Praktische Naturphilosophie fuer die Umweltpolitik, Munich: Carl Hanser Verlag 1984, p. 24.
12 See also G. Patzig, "Oekologische Ethik – innerhalb der Grenzen bloßer Vernunft" (1983), in: G. Patzig, Gesammelte Schriften, vol. II, Goettingen: Wallstein Verlag 1993, p. 180.

This is why philosophy has traditionally seen our responsibility towards nature (for its own sake) as a question that concerns man's responsibility *towards himself*. Immanuel Kant's view, for example, is that in

> beholding what is *beautiful* but inanimate in nature (…) a tendency to mere destruction (…) is contradictory to man's obligation to himself because it weakens or eliminates that feeling in man which is not in itself already moral, but which nevertheless at least prepares that mood of sensuality which promotes morality: namely, to love something without the intention of use.[13]

In this context, Kant particularly mentions the suffering of animals as a reason for moral responsibility and obligations towards nature. When man inflicts suffering on animals, his 'sympathy' for their suffering declines, so that 'a natural disposition which facilitates morality in relation to other people weakens and is exterminated little by little.'[14] This means that man's obligations towards nature are obligations towards ourselves. Endangering nature, in Kant's argumentation, threatens morality. Not because nature is itself a (moral) subject, but because, in this sense, the interaction with nature also falls under moral categories.

However, one judges this type of argument, this much is clear (as it is in Kant): (rational) nature-based ethics (*Naturethik*) cannot be grounded 'in itself', i. e., it cannot be derived 'from the perspective of nature,' which would thus be understood as a moral subject. Nature-based ethics can only be part of ethics in which (rational) man is the subject. This is the real core of what is called *anthropocentricity* in ethics. Accordingly, reason-based *ethics* (*Vernunftethik*) encompasses nature-based ethics (*Naturethik*), be it for reasons that Kant mentions or for reasons that I have discussed here. To the extent that rational action begins to change nature in a way that endangers 'natural' and 'reasoning' nature, we can say that nature is ethical, and not just in a physical or economic sense becomes the limit of our actions. No other rationalities can set this limit, but neither can nature, i. e., a nature that 'opens its eyes.'[15] As in any enlightened ethics, i. e., ethics based on the view of reason, this limit is the limit of reason (*Vernunftgrenze*).

Even if we incorporate nature-based ethics into reason-based ethics, we still cannot be sure what nature is and what it should be. For technological cultures with their appropriating structures, the answer to such questions is basically a mystery. In a certain sense, this is even true of an *ecological* perspective, from which one would most likely expect the correct answer. Ecological judgements, too, cannot say what a non-appropriated and non-cultivated nature should be. Nor can they say how we should behave

---

13 I. Kant, Die Metaphysik der Sitten: Tugendlehre § 17, Werke in sechs Baenden, ed. W. Weischedel, Darmstadt: Wissenschaftliche Buchgesellschaft 1956–1964, vol. IV, p. 578.
14 Ibid., p. 579.
15 J. Habermas, "Technik und Wissenschaft als 'Ideologie'," in: J. Habermas, Technik und Wissenschaft als 'Ideologie', Frankfurt: Suhrkamp 1968, p. 19.

adequately towards nature. And just as medicine cannot express how healthily we should live, but only how we should live once we know how healthy we want to be; so, ecology can also not say how natural nature should be, but only what nature is relative to our ends and how we achieve them. In other words, when ecology has to decide what the proper relationship between technological culture and nature is, it, too, pre-supposes a particular conception of what nature should be.

3.   Ecology does not develop these conceptions by itself

This can be illustrated by using the central concept of ecological balance, which in ecology is primarily a descriptive concept. It defines the state of an ecosystem in terms of a fluid equilibrium. In general discussions, however, it is often used (and misused) as a prescriptive (normative) concept to describe a state that should not be disturbed for its own sake.[16] Besides, it is not preserving a state of equilibrium that shows the very essence of nature but its dynamical renewal. Chemical equilibria, for example, would mean the end of all life.[17] Therefore, it does not mean retaining the nature and thus the environment in their present state once and for all and at all costs. A 'standardized' environment, clearly defined by environmental standards, is a strange idea. Nature has been experimenting every day for millions of years, 24 hours a day. Its laboratory is everywhere; even in our laboratories, its lead is unassailable and will likely remain, notwithstanding what has been said about a cultural nature. Environmental standards aim to reduce environmental threats in this dynamic process; we will almost always be lagging behind natural evolution.

That ecology itself cannot develop conceptions of what nature should be can also be illustrated by indicating the grounding of ecology in man's (rational) self-interest. For example, protecting endangered species:

> In answering the question which of the many species – apart from moral or aesthetic con-
> siderations – we will need most in the future since they could be potentially useful to us
> later, the best answer is the least precise: as many as possible. The reason is simply that all
> species are gene banks that maintain themselves for us in the least expensive way (…).
> This, of course, is especially true of the primary chemical producers, the hundreds of thou-

---

16 See P. W. Taylor, Respect for Nature: A Theory of Environmental Ethics, Princeton N. J.: Princeton University Press 1986, p. 299.
17 See R. Lewin, "In Ecology, Change Brings Stability," Science 234 (1986), p. 1072 ("ecosystems are rarely if ever at equilibrium. They are continually being perturbed – particularly by shifts in the environment – and are therefore in a permanently unstable state, constantly seeking an equilibrium that is always moving. (…) 'If they ever did come to equilibrium I don't think we would like them very much. The reason is that, in coming to equilibrium, the rich ecosystems we see today would inevitably lose many of their species.'").

> sands of plant species. (…) This is why our first and greatest care, for mere utilitarian reasons, should be the preservation of as many plant species as possible.[18]

This is how things are, even if it pleases neither the philosopher among the ecologists nor the teleologist among the philosophers.

Thus, ecological steps to a proper way of dealing with nature are also not simple nor straightforward. They generally range from a *rule of prudence* (limiting exploitation and dealing with damages), in which case they are themselves economically oriented, to moral commandments (responsibility for nature). In this respect, they share problems with ethical orientations. There is also the attempt to develop a 'philosophical' way of looking at things in which teleological and evolutionary points of view constitute a mode of argumentation that is difficult to understand. The following conviction, for example, reveals this:

> (a) The belief that humans are members of the Earth's Community of Life in the same sense and on the same terms in which other living things are members of that Community. (b) The belief that the human species, along with all other species, are integral elements in a system of interdependence such that the survival of each living thing, as well as its chances of faring well or poorly, is determined not only by the physical conditions of its environment. It is also determined by its relations to other living things. (c) The belief that all organisms are teleological centres of life in the sense that each is a unique individual pursuing its own good in its own way. (d) The belief that humans are not inherently superior to other living things.[19]

What is presented here as a 'biocentric' point of view follows the already mentioned 'teleological' idea of replacing rational ethics (*Vernunftethik*) with nature-based ethics (*Naturethik*, 'biocentric' ethics).

With that, returning to the boundaries between an anthropocentric and an ecological (or, as it is also labelled, physiocentric) approach concerning nature and normativity. Anthropocentrism means, in short, and perhaps a bit misleading, that man is the measure of all things; ecocentrism or physiocentrism implies that nature is the measure of all things. From the anthropocentric perspective, in matters of ethics and nature, it is only possible to argue from man's perspective; nature has no moral value of its own. From the ecocentric or physiocentric perspective, by contrast, nature has such own (absolute) value, which, at the same time, determines man's duties to nature. More precisely, in the case of *physiocentrism*, further distinctions are made between *pathocentrism* (all sentient beings have a moral value), *biocentrism* (all living beings have a moral value), and *radical physiocentrism*, which makes the whole of nature the carrier of moral

---

18 H. Markl, Natur als Kulturaufgabe, pp. 337–338.
19 P. W. Taylor, Respect for Nature: A Theory of Environmental Ethics, pp. 99–100.

values.[20] What all these variations have in common is that values, which are indeed the result of valuations, are declared to be parts of nature itself.

With the idea of nature-based ethics as biocentric or physiocentric ethics, ethics does not just become dependent on a particular world view, in this case, the conviction mentioned before. There is also the threat of a new *biologism*, namely the grounding of an ethical orientation in biological facts. Here, ethics becomes a part of the natural history of man. Questions like the following point in this direction (quoting from the biologist Hans Mohr):

> Is our inability to react with reason to the challenges of our times a cultural mis-adaption that can be (easily?) corrected, or is it the expression of an evolutionary inheritance that proves to be a mis-adaption in the present cultural situation. Are there biological limits which block man's path to eternal peace and global ecological reason?[21]

If the second of these options were true, man would be the victim of his own (biological) nature, enlightenment – and hence also reason-based ethics (*Vernunftethik*) – would remain stuck in a biological dead end. Further suggestions contain little comfort:

> Man is a product of evolution, so his cognitive dispositions too (patterns of thought and patterns of knowledge) must originate in the course of Darwinian evolution. Our emotional patterns of knowledge correspond to the real world precisely because they evolved in adaptation to this world in the course of human evolution. How we think, and the range of our thoughts are anchored in our genes.[22]

This means ethics arises out of a (problematical) evolutionary epistemology. Behind our moral deficits stands a biological evolution that is not completed and cannot deal with us – with its inborn patterns of behaviour and action: The

> human mind, adapted to the middle dimension of the Pleistocene Era (the age of gatherers and hunters) was not created to understand the modern world and its dangers.

> The pathological thoughtlessness with which we foolishly reproduce ourselves and completely exploit the planet is our biological inheritance. We do not realize what we are doing.[23]

---

20 See A. Krebs, "Oekologische Ethik I: Grundlagen und Grundbegriffe," in: J. NidaRümelin (Ed.), Angewandte Ethik: Die Bereichsethiken und ihre theoretische Fundierung. Ein Handbuch, Stuttgart: Alfred Kroener Verlag 1996, pp. 346–385. Furthermore A. Krebs, Ethics of Nature: A Map, Berlin and New York: Walter de Gruyter 1999.
21 H. Mohr, "Evolutionaere Ethik," in: Information Philosophie 4 (1986), pp. 4–5.
22 Ibid., pp. 5–6.
23 Ibid., p. 8.

This is probably the case. It is precisely rational cultures that are often, with respect to their appropriation patterns, struck by irrational blindness. But is biology here really an explanation or even an excuse? The way the author rounds off his ideas sounds more plausible:

> Our future will depend on whether and to what extent man succeeds in his actions in breaking away from the obsolete fitness maxims of biological evolution and creating the decision-making freedom he needs for a good life.[24]

In fact, that is man's actual task as a rational being. Attempts to view the rationality of this being either an evolutionary product or as running contrary to evolution lead here in the wrong direction. What is at stake is not the playing of man's nature, including his rational nature, against biological evolution (or in favour of it), but making nature the expression of reason in man and his culture. Reason-based ethics (*Vernunftethik*), however, does precisely this. Nature-based ethics (*Naturethik*), that teleologically or in whatever other way going beyond the limits of reason-based ethics, does not. In other words, replacing anthropocentric or reason-based ethics with ecological or physiocentric ethics returns to a speculative philosophy of nature and systematically into a blind alley. The future of ecological reason lies not in any kind of physiocentrism but in a reason-based, normative or prescriptive ethics, freed of all elements of destructive human selfishness and ecological romanticism.

---

24 Ibid., p. 15.

Opening Ceremony 2018

ELENA WIEDERHOLD

September 19, 2018
Building 'Normative Orders', Campus Westend
Goethe University Frankfurt

In a ceremonial act with science, politics and university guests, the Johanna Quandt Young Academy (JQYA) at Goethe University was opened on September 19, 2018. Prof. Matthias Lutz-Bachmann, one of the two founding directors of the JQYA and the creative mind behind the idea, highlighted: 'The Academy's aim is to bring very outstanding young scientists from all over the world to Goethe University so that they can work here in cooperation with the best researchers.' The establishment of the JQYA was made possible by generous financial support from the Johanna Quandt University Foundation.

The second founding director of the JQYA, Prof. Enrico Schleiff, emphasised the importance of an academy for young scientists at the opening ceremony:

> I owe my enthusiasm for working in an academy to the Volkswagen Foundation, which let me work in a similar context, and I would like to pass on the positive experiences to young colleagues. Especially the interdisciplinary exchange and breaking out of one's specialist fields is a precious and formative experience that allows a completely different view of one's own research and field of research. I am deeply convinced that the young colleagues will enjoy the academy work!

Prof. Anderl, the President of the Academy of Sciences and Literature Mainz, spoke in his welcoming speech about the importance of funding for postdocs and how critical the postdoc phase is in career development. In particular, Prof. Anderl accentuated the differences and importance of funding through an academy compared to general funding (e. g., through post-graduate schools). An academy is an educational institution, in his view, that promotes the exchange of ideas and intellectual flow and forms a scientific community that strengthens itself through the exchange of ideas. Following his speech, Prof. Anderl reiterated the desire for cooperation, and emphasised the relevance of the JQ Young Academy for the Rhine-Main Alliance cooperation.

JQ Young Academy welcomed in 2018: Eva Buddeberg (Political Sciences), Sandra Eckert (Social Sciences), Daniel Merk (Pharmaceutical Chemistry), Hanieh Saeedi (Marine Biology), Peter Smith (Linguistics) and Florian Sprenger (Media and Cultural Studies) as JQYA fellows; Federico L. G. Faroldi (Philosophy, Ghent University) as international fellow and Jasmin Hefendehl (Neurobiology) as member.

The JQYA, together with the Goethe University Presidential Board, honoured the first cohort of Distinguished Senior Scientists 2018–2020: Prof. Seyla Benhabib (Yale University, Politics and Philosophy), Prof. Gunnar von Heijne (Stockholm University, Biochemistry) and Prof. Nicola Spaldin (ETH Zurich, Physics).

Discipline-Related Debates

ELENA WIEDERHOLD

Criticism of Religion in the Enlightenment Era
International Workshop on February 27, 2020
Goethe University in cooperation with 'Normative Orders'

Organiser: Eva Buddeberg
Guests: Luke Collinson (London), Ieva Motuzaite (Berlin), Herbert de Vriese (Antwerp), Anna Tomaszewska (Krakow) and Dagmar Comtesse (Frankfurt).

Eva Buddeberg contributed to the scientific discussions on the Academy's platform by reflecting on the concept of religion within the discipline of philosophy. The need for such a debate was that, in recent years, the close coupling between the concepts of modernity and secularisation has loosened. At the same time, the interest in the relationship between religion and the Enlightenment has been growing, and so has the interest in academia in critically examining the concept of an 'enlightened Christianity': What is the record of the Enlightenment? What has been achieved? What, historically speaking, was its programme of religious criticism? What insights can still be gained from the approaches of that time? And what significance can we still ascribe to them today? These questions are relevant not only from the perspective of theology and political philosophy but also regarding modern societies and their challenges of coping with religious plurality, including the difference between orthodox forms of religion and increasingly free-floating religiosity. The participants of the workshop seek to find some answers to these pressing questions by analysing and discussing the history of enlightenment with regard to its criticism of religion in general and Christianity in particular.

Interdisciplinary Tandem-Symposium[1]
October 1–2, 2019
Representation of the State Hessen to the European Union, Brussels

The first academic year of the Johanna Quandt Young Academy, which was themed 'Nature and Normativity', culminated with a two-day symposium at the Hessian Representation in Brussels. The symposium provided a platform for reciprocal discourse between JQYA fellows/members, international guest speakers, representatives of the European Parliament, the European Commission, the European Research Council and non-profit organisations.

The symposium was divided into six panel discussions, each with different issues relating to an aspect of the theme 'Nature and Normativity'. The particularly dynamic and interactive nature of the symposium came from the fact that all panels were conceived on a focus topic of the respective fellow. Their challenge was to moderate a panel which was not their own, but rather one of a colleague's from another discipline. This also provided a wider outlook on other disciplines and a broader line of questioning from the perspective of one's own discipline.

Panel organiser: Federico Faroldi

The symposium opened with panel I 'Natural and Normative Properties', which explored the evaluative properties of 'natural' and 'beauty' in scientific theories. A physicist and a philosopher led a discussion on the different criteria required, if any, to distinguish between natural, non-natural and normative properties. Two central questions were how and why the seemingly evaluative properties, such as 'naturalness' and 'elegance', are used in scientific theories.

Panel organiser: Eva Buddeberg

The concept of 'naturalness' was echoed in the next panel (II) in a central question of how the 'natural' becomes the 'normative' idea of human technology, arts and behaviour. The panel also discussed how the paragon of 'the natural' resembles or differs from the ideal of biomedical research in the 21st century when addressing the concept of 'motion' in biology as a definition of life. If an artificial 'organism' implanted into a living form is designed to autonomously conduct a living function in that form, is this considered 'natural'? Somewhat akin to how the 'ideal of human behaviour in the

---

1 Kindly supported by the Representation of the State Hessen to the European Union.

18th century' once perceived as natural is no longer considered to have been so. More-over, how does the natural differ from the nature? To what extent does nature repre-sent the obstacle to be overcome and, at the same time, the normative ideal? How can the natural be artificially created or learned and yet remain natural?

Panel organiser: Peter W. Smith

Both a linguist and a neurobiologist presented their research in panel III to exchange ideas on the nature of language and language diversity. The central question was why human beings seem to have a 'structured' language system (with grammar etc.) while animals do not. What are the underlining biological and neurological differences? When and how did the 'biology' change? Based on the presentations, new impetus for rethinking language parameters and the structure of universal grammar was discussed, and the central question of biological and neurological evidence for the possible evo-lution of language was examined.

Panel organisers: Daniel Merk, Hanieh Saeedi

Moving from language to information and databases, panel IV addressed the varying expectations that scientists from different disciplines have of global scientific data-bases. Scientists from UNESCO's Deep Sea Node, and from the fields of linguistics and pharmacy, led a discussion which expounded the increasingly important role that scientific databases play in the era of globalisation and open science to ensure world-wide information sharing. Database requirements, different formats, best practices for processing and presenting available data for research and future developments were discussed. Also addressed were questions such as: how can we ensure a sustainable and consistent quality of data? What improvements do we expect and desire? The partici-pants discussed their responsibility as scientists to make the data available for broader use, following a FAIR data principle to make the data Findable, Accessible, Interoper-able, and Reusable. For instance, the FAIR data management in OBIS UNESCO aims to ensure sustainable research marine data promoting high-quality research.

Panel organisers: Federico Faroldi, Christian Münch

These divergent expectations arising from a multidisciplinary approach was followed by panel V on 'Disciplinary and Interdisciplinary Research' with ERC and HORI-ZON Programme representatives, and started with questions about chances, risks and trends in disciplinary versus interdisciplinary research. When do complex societal problems require an interdisciplinary approach? Two researchers and two policymakers were invited to share and compare their professional perspectives on contemporary and future interdisciplinary research. The focus was on needs, hopes, expectations and frustrations regarding the organisation and results of interdisciplinary research. A lively debate between the panel and the audience ensued, on how the interdisciplinary nature of a research proposal and the results of an interdisciplinary research project should be better evaluated in future. Young scientists see interdisciplinarity not only as a great opportunity to collaborate across disciplinary boundaries, they also recognise the higher risks of failure that it carries and the challenges it presents in teaching. This take-home message was delivered to the policymakers to improve future developments in science policy.

Panel organiser: Sandra Eckert

In the final panel (VI), the challenges posed by plastics in the EU's transition towards a circular economy was the topic of discussion. The concept of a circular economy was introduced and defined from the perspective of social sciences, policymakers, NGOs and business, and current policy practices in the European context were discussed. While recent policy measures have targeted plastics in particular, circularity poses a challenge to a variety of industry sectors such as the European paper industry. The panel participants elaborated on these challenges and identified the need for further action from the incoming Commission.

In addition to the interdisciplinary panel discussions, the participants had sufficient scope for discipline-related discussions. At the conclusion of the first day, participants were also given the opportunity to meet and speak with the winners of the Peace Prize of the German Book Trade 2018 (Friedenspreis des Deutschen Buchhandels 2018), the German Professors of Cultural Studies, Aleida and Jan Assmann.

Invited panel members:

Prof. Ralf M. Bader
Institute for Future Studies in Stockholm. *Ethics, metaphysics, Kant, political philosophy and decision theory*

Prof. Theresa Biberauer
University of Cambridge. *Bio-linguistic, naturally acquired languages and forms of languages shaped by conscious human intervention*

Dr Jan von Brevern
Freie Universität Berlin. *Visual media and photography, history of aesthetics, cultural history of nature*

Dr Sabine Hossenfelder
Frankfurter Institute for Advanced Studies. *Theoretical physics*

Dr Francesca Grisoni
ETH Zurich. *Pharmaceutics, artificial intelligence and computer-assisted drug design*

Dr Markus Kunze
University of Vienna. *Biology, biochemistry, brain research and evolution*

Ulrich Leberle
Raw Material Director, Confederation of European Paper Industries

Delphine Lévi Alvarès
European coordinator of the 'Break Free From Plastic' movement and Zero Waste Europe

Julie Mennes
Ghent University. *Cross-disciplinary collaboration through philosophical and computational linguistic tools*

Dr Alice Rajewsky
Head of Sector Social Sciences and Humanities at the European Research Council Executive Agency

Prof. Kenneth Safir
Rutgers University (USA). *Linguistics, pioneering the development of databases in linguistics research*

Axel Singhofen
The Greens/European Free Alliance in the European Parliament. Health and environment policy adviser

Renzo Tomellini
Head of Unit HORIZON Strategic Planning and Programming, Head of Unit Mission and Partnerships at the Directorate-general for Research and Innovation of the European Commission

Dr Janina Wellmann
Leuphana University Lüneburg. *History and epistemology of life sciences in the modern era*

III.   2019–2021 Tenets, Variations, Transformations

As basic principles, **tenets** guide the practice of scientists across disciplines as well as the social and political life in our societies. Even though widely accepted, such principles are, however, subject to important **variations**. Any changes over time that are **transformative** in scope challenge the fundamental nature of tenets. Such paradigm shifts result in a conceptual re-thinking such as discovering unknown natural laws, changes of institutional policies, novel findings, the definition of novel traditions, priorities, research fields, etc.

The three key terms of tenets, variations and transformations offer multiple avenues for a genuinely interdisciplinary discussion by posing scientific questions such as:

–   What are the 'tenets' we as a community adhere to in various contexts such as scientific practice across disciplines or in society and politics?

–   What type of 'variations' can we discern from these tenets, and under what conditions are these still acceptable deviations of a given norm?

–   How do we define 'transformations', and to what extent do these deconstruct or replace pre-existing tenets and their variations?

–   How are the assumptions of these concepts correlated in the different disciplines?

Keynote Lecture 'Intellectual Authority in Classical Antiquity'

HARTMUT LEPPIN

When I accepted the invitation to give the keynote lecture for this event and suggested my topic to Matthias Lutz-Bachmann, I did not expect it to be as topical as it proved to be. But even in those distant days before Covid-19, it was topical. This is because one of the most dampening experiences of recent years, particularly from the point of view of our academic world, has been the loss of authority of the rational argument, as demonstrated, for example, by this rejection of expert opinion by a highly educated British politician, Michael Gove, in the run-up to the Brexit vote. Experts were accused of speaking on the basis of their own interests, of representing only a narrow elite, of often missing the right outcome and of having no sense of what really mattered. They are described as being far removed from what is then defined as the people, who are assumed to have a natural, general feeling for the right thing, which does not need any justification. In addition, we observe an emotionalisation of politics, a formulation of dangerous, inflammatory political statements in Twitter format. I will not mention the name of the person we are all thinking of. This criticism of expertise is rapidly expanding into anti-intellectualism in a broader sense.

During the Covid-19 pandemic, however, we have experienced a dramatic change. A certain variant of intellectual authority, academic expertise, has taken on a new clout. With the outbreak of the pandemic, epidemiologists, expert scientists, became public figures. Through articles, interviews, but also podcasts, they have gained great public attention. One tabloid newspaper even called on readers to vote for their favourite epidemiologist. And of course, it was no disadvantage that an epidemiologist such as Christian Drosten looked like everybody would imagine: a good-looking scientist, who is especially attractive with uncombed hair. The non-academic public quickly learned that even science cannot provide reliable predictions in complex cases, they were prepared for a certain time to listen to differentiated argumentation, to respect the uncertainty of scientific knowledge and most of them followed the advice of the experts. Remarkably, the experts who were accepted by the respective government gained prestige, even if they gave very different advice – in Sweden, state epidemiologist Anders Tegnell became a popular figure. Things were more convoluted in Great Britain, but I cannot discuss that here. What matters is that trust in experts was closely intertwined with trust in governments. Obviously, it is not just a question of the power of the argument, but of the assurance that political action is based on arguments.

Things have changed in the last few weeks. Germany's most famous epidemiologist has been exposed to a smear campaign, with protests in many countries against so-called corona measures, which unfortunately were quickly hijacked by extremists and crackpots. Nevertheless, the problem that is articulated in these manifestations is important. It must not be left to experts alone. Both the measures, which, after all, entail considerable restrictions on freedom, and the influence of the experts have to be closely monitored in a democracy.

What role may experts play in a democracy? What degree of influence is legitimate? Doesn't the influence of the elitist group of experts pose a threat to democracy, which is based on equality? I found it most impressive that very few experts succumbed to the temptation of confusing their role with that of politicians, even if parts of the media expected otherwise. Politics necessarily takes place in a space of insufficient information; whereas scientists must weigh up probabilities, politicians have to make decisions without ever having the certainty that they are the right ones. They have to take different factors into account.

What is the importance of health protection, which is a priority for epidemiologists, compared with other factors, education, economics, fundamental rights? How long will which restrictions be accepted by the public? This is where completely different experts come into play, and also non-experts, even moods have a significant influence. Currently, we are witnessing these debates gaining momentum.

It is not only because I am a historian that I find it important to look back at such situations and to shed light on the role of intellectual authority in other epochs. By intellectual authority, I mean a form of authority that is not based on status, origin, wealth or spirituality, but on the power of the argument that is detachable from the person, that is valid as such in the view of the respective contemporaries. The bearer of intellectual authority can use it particularly skillfully, but unlike the wise man, he is not attributed authority that lies in the person. Rather, he must always prove himself, even reckon with counterarguments from people who rank below him. It goes without saying that this is an ideal type; in reality, various forms of authority are combined.

I will discuss the subject using the example of the first democracy in world history, that of classical Athens, in the 5th century BC, 2500 years ago. The Athenians did not have democracy before them, nor did they aspire to democracy. Their political order was not the result of purposeful reforms guided by an idea of democracy, but rather a consequence of aristocratic rivalries, military challenges and contingent circumstances. Contingent was, for example, the fact that Persian kings thought they had to conquer Greece. As a result, the Athenians had to recruit more citizens for military service, specifically naval service, than ever before. The Greek citizen soldiers were victorious against the apparently superior Persians. But those who were militarily essential, the men who rowed the ships, could also strive for more power in politics. The citizen soldiers were able to claim influence over the nobility, and the People's Assembly became the main deciding body in the democratic city.

In addition, ordinary Athenian citizens took up increasingly more public functions. Only in exceptional cases were civil servants elected, in most cases by sortition. Thus, every male citizen could sit on the council that discussed the agenda of the People's Assembly, every man could become a judge, every man could be in charge of the wells or of the sale of confiscated goods, etc. Since offices were only held for one year and one could not sit on the council for more than one year, and the courts were always drawn by lot, every Athenian had the chance to take up a responsible position in public life several times in his life. The tenet behind this was that every citizen was equally competent to do politics, an absolutely unusual idea in antiquity.

The Athenian aristocrats did not act as a closed group, but individually, nor did they have a fixed clientele on whom they could rely. Those who wanted to assert themselves in the People's Assembly could not rely on personal authority and could not bring any superior power into play; economic resources could hardly be activated either. One had to argue, and thus intellectual authority gained importance.

In the classical tradition, Pericles was the epitome of the new, clever type of politician, who was linked to many great minds of the time, such as the sculptor Phidias and the tragic poet Sophocles. Pericles is said to have placed himself entirely at the service of the city, abruptly ending all friendships when he went into politics. He held the most important electoral office in the city for several years, from 443 to 429, because he was one of the 10 so-called strategoi who led the army, but who also had decisive influence on domestic politics. The great historian Thucydides characterized him as follows:

> Pericles indeed, by his rank, ability, and known integrity, was enabled to exercise an independent control over the multitude – in short, to lead them instead of being led by them; for as he never sought power by improper means, he was never compelled to flatter them, but, on the contrary, enjoyed so high an estimation that he could afford to anger them by contradiction. Whenever he saw them unseasonably and insolently elated, he would with a word reduce them to alarm; on the other hand, if they fell victims to a panic, he could at once restore them to confidence. In short, what was nominally a democracy became in his hands, government by the first citizen. With his successors it was different. More on a level with one another, and each grasping at supremacy, they ended by committing even the conduct of state affairs to the whims of the multitude.[1]

---

1 Thucydides, 2.65.8–10, transl. by Richard Crowley.

The implicit theory here is that a democracy can only function if the leading politician has a high intellectual authority, but this has to be combined with personal integrity. Pericles, as Thucydides knew, was of noble birth. Nevertheless, typically intellectual authority would weaken traditional elites: if compared to the social and economic power of the aristocrats, it is potentially much more broadly based in terms of social origin and opens up new access possibilities for non-aristocrats. It can be acquired and is not hereditary. Moreover, it must always prove itself anew; it is not enough to refer to past merits of one person or even to ancestors. Thus, it seems to be closely linked to democracy. For in its central organ, the People's Assembly, in which all citizens were allowed to participate, every speaker had to argue in a convincing way. Here the argument prevailed, not the person, at least the best presented argument. Everyone had to face the discussions in the People's Assembly, even the famous Pericles. He had to be re-elected year after year, he had to campaign for his politics over and over again, and failed repeatedly.

Everyone had the same right to speak in the popular assembly or more precisely, every male Athenian citizen. Athens' democracy excluded women and accepted the existence of slaves. It was a democracy, but it was not based on the idea of universal human rights. Rather, it was based on the right of citizens, who were supposed to be equal.

But things were not so simple after all. For one thing, the accessibility of education still depended on the financial means and the availability of young people. You had to have the money and time to go to school. It was a society based on the idea of political equality, but it was a society of social inequality.

The Sophists, for example, who offered various forms of education, including dialectics, taught such things. To this day, the reputation of the Sophists suffers greatly from the fact that Plato condemned them as people who were not interested in the truth, but only in making money. Nowadays historians are much more cautious in judging the Sophists. They took money for teaching, that's true. But the criticism of making money is somewhat cheap from the mouth of a well-to-do man like Plato.

Above all, the essential importance of the Sophists for democracy has become clear, precisely because they taught rhetoric, which they treated not as a personal characteristic, but as a skill that could be trained, and which allowed even people who did not come from noble families to participate effectively in political debates. It was, however, not egalitarian. The inequality merely shifted. Only wealthy families had the financial resources to provide their children with a good education. The goal of education was the success of those who used rhetoric. They should learn to work out the plausible, the eikós. Although rhetoric has a high emancipatory potential, it is primarily about shifting elites. The increase in the importance of intellectual authority opens up a society but means exclusion in other areas. That is why the distrust of intellectuals then and now should not be condemned in advance. It is a genuinely democratic reflex of less privileged groups that emphasises the equality of all.

In Athens, too, criticism of the rhetoricians soon came up. This too is reflected in the historical work of Thucydides, in an episode set in 427 BC, during the long Peloponnesian War between Athens and Sparta, which lasted from 431 to 404 and ended with the defeat of Athens. During this war, the city of Mytilene, an important ally of Athens, joined up with Sparta, but was conquered by the Athenians. They decided on a cruel verdict: all adult men should be killed, women and children enslaved. Immediately a ship sailed to Mytilene to convey the order. But the next day the Athenians started wavering and discussed the issue in another popular assembly. There also Cleon rose to speak. He is characterised by Thucydides as a politician especially recognised by the people. Nevertheless, he indulged in a veritable critique of democracy. The Athenians were naïve, he said, and wrongly believed that a friendly approach to foreign policy would be rewarded; they overlooked the fact that their so-called allies hated them. Then he continues:

> The most alarming feature in the case is the constant change of measures with which we appear to be threatened, and our seeming ignorance of the fact that bad laws which are never changed are better for a city than good ones that have no authority; that unlearned loyalty is more serviceable than quick-witted insubordination; and that ordinary men (i. e. regular people) usually manage public affairs better than their more gifted fellows.

> The latter are always wanting to appear wiser than the laws, and to overrule every proposition brought forward, thinking that they cannot show their wit in more important matters, and by such behavior too often ruin their country; while those who mistrust their own cleverness are content to be less learned than the laws, and less able to pick holes in the speech of a good speaker; and being fair judges rather than rival athletes, generally conduct affairs successfully. These we ought to imitate, instead of being led on by cleverness and intellectual rivalry to advise your people against our real opinions.[2]

The following paragraphs of Cleon's speech are also interspersed with polemics against fine-sounding speeches. He demands infallible consistency of behaviour based on the laws that exist. Reflection destabilises societies; for those who pretend to think more thoroughly are only concerned with cultivating their image. Many statements ascribed to Cleon are reminiscent of the anti-intellectualism of modern populists – with the difference that Cleon values law highly. But in Athens, unlike in modern times, the law is not subject to the interpretation of experts but is a matter for the people. In this case, Cleon could not assert his position, but mostly he was a successful politician, particularly with this populist gesture. Unlike the previous politicians like Pericles who appeared more measured, he roared around and moved violently, again demonstrating that he was not an elitist man.

---

2 Thucydides, 3.37.3–5; transl. by Richard Crowley.

The philosopher Aristotle, or one of his students, commented:

> The head of the People was Cleon …, who is thought to have done the most to corrupt
> the people by his impetuous outbursts, and was the first person to use bawling and abuse
> on the platform, and to gird up his cloak before making a public speech, all other persons
> speaking in orderly fashion. [3]

According to this, Cleon not only attuned to the pulse of the people and converts it,
but also contributes to their ruin with his impulsiveness. In his deliberately uncouth
manner and his ruthlessness he reminds me of someone I will not name. He thus became the representative of an aggressive war policy that was to plunge Athens into
the abyss. Others also contributed to the fact that during the war the Athenians tore
themselves apart in fierce internal conflicts, and sometimes made flagrant mistakes,
such as a campaign to Sicily, which was said to be easy to win but which cost the lives
of thousands of Athenian soldiers.

These events show the ambivalence of rhetoric in democracy. Democracy was its
breeding ground and at the same time it could endanger democracy. Many Athenians
believed that Athens had lost the war because of the rhetoric. They lived under the
impression that the rhetoricians were interested in winning the dispute and not in the
truth or the well-being of the city. Therefore, they seduced the people and served their
own fame rather than the city. Thucydides, on his part, asserted in the chapter quoted above that the Athenians could have won if Pericles had guided them further, for
he had correctly calculated everything. He therefore trusted that victory was possible
even in a democracy.

For the people, the demos, was actually considered good and unfailing in principle, every citizen was seen as competent for almost everything political. Even some
philosophers trusted in what is called *Schwarmintelligenz* in modern German: the people as a whole will make the right decisions unless they decide too quickly or are influenced by the wrong people. In the end the people as a whole are more intelligent than
any individual, than any expert. This was a crucial tenet of democratic thinking.

But how should one prevent the people from being seduced? How should the Athenians react to the fact that time and again officials and speakers had cheated them and
that the people, who are actually infallible, have therefore made the wrong decisions?
How could Athens be made great again?

One possibility was to reject democracy, as Plato did. Plato is an author whose spell
is hard to escape. The philosopher disliked the democratic order but benefited from
the stability of the city and its relative prosperity. He did not, however, give his advice
to his fellow citizens, at least not where they sought it, in councils and popular assem-
blies. Instead, he held talks in his private university, the Akademía, which lent its name

---

3 Aristotle, 28.3–4; transl. by H. Rackham

to many future academies. And since you are an academy, let me briefly remind you of the beginning. The academy was originally a cult association for the not particularly important hero Akademos, just outside the city. Since there was no general association law, the formation of a cult association was a way to get organised. I am convinced that the academics not only discussed important issues, but also made sacrifices to the divine Akademos. This might have been part of the community life that turned away from the city.

Plato was related to opponents of the Athenian democracy. From the point of view of most Athenians, he was therefore one of the traitors. Many of you will know that he was a disciple of Socrates, who was sentenced to death in 399 because he was accused of introducing new gods and corrupting youth. For someone like Plato, this was certainly a sign of the failure of the courts, which were made up of ordinary citizens.

Plato withdrew from the polis. He does not appear in public documents of the time. Apparently, no one cared about what this man, who held his profound conversations in the grove of the Akademos, thought about day-to-day political affairs. Did his contemporaries take him seriously? Was he a morose madman for them? It's hard to say. No one could have guessed what kind of *Nachleben* this thinker on the margins of the city would have.

He, for his part, was a keen observer of democracy. This is evident in many parts of his oeuvre, which consists mainly of dialogues. In other words, Plato has various historical figures discussing important themes, usually with his teacher, Socrates. These are not protocols of actual conversation, but fictions. Plato – not the only one – avoided the style of philosophical treatise with which we are familiar. Galileo Galilei still deemed it advisable to write a dialogue.

One of the most famous dialogues discusses the role of rhetoric. It is named after a Sicilian orator, Gorgias, who is said to have brought the art of rhetoric to Athens during the Peloponnesian War and who had many students. Here too, it is not about the historical Gorgias, but about the image of him and his students. Socrates, Plato's Socrates, asks Gorgias what the art of rhetoric is:

> Since you claim to be skilled in rhetorical art, and to be able to make anyone else a rhetorician, tell me with what particular thing rhetoric is concerned: as, for example, weaving is concerned with the manufacture of clothes, is it not?
>
> Gorgias: Yes.
>
> Socrates: And music, likewise, with the making of tunes?
>
> Gorgias: Yes.
>
> Socrates: Upon my word, Gorgias, I do admire your answers! You make them as brief as they well can be.
>
> Gorgias: Yes, Socrates, I consider myself a very fair hand at that.

> Socrates: You are right there. Come now, answer me in the same way about rhetoric: with what particular thing is its skill concerned?
>
> Gorgias: With speech.[4]

Socrates can quickly show that this statement is not enough, since many other arts, such as the healing arts, are also concerned with words. Finally, he tempts Gorgias to say that rhetoric is about the greatest and best things.[5] But Socrates demands further clarification by asking what is the greatest good, and here a political aspect comes into play:

> Gorgias: I call it the ability to persuade with speeches either judges in the law courts or statesmen in the council-chamber or the commons in the assembly or an audience at any other meeting that may be held on public affairs.[6]

But Socrates wants to know what the contention is all about, because other arts are also convincing or persuading. The Greek word peíthein shimmers between the two terms. Gorgias explains that it is about what is just and unjust.[7] Socrates, however, can prove to him that he is at best able to produce a belief, but no knowledge of what is just and unjust. It is easier to convince an incompetent audience than a competent one. Yet Gorgias even boasts that he does not need expert knowledge. But for Socrates the question of the relationship between rhetoric and justice remains open.

Now Gorgias' student Polus jumps into the breach and turns the tables on Socrates, asking him what rhetoric is for him. The philosopher provocatively puts it on a par with cooking and styling, which are concerned solely with the production of pleasing things; it is nothing but a form of flattery.[8] In consequence, he accuses democratic politicians of blandishment. Even Pericles does not come off well.

Speakers, Socrates says, have no power in the polis, nor do the tyrants. For in the end, they are flatterers of the people and not able to act in the spirit of good. In what has been said, the claim of Platonic philosophy as compared to rhetoric is clearly evident: philosophy acts in the sense of truth and good; rhetoric, on the other hand, is no more than an instrument that can be used for cheating. Plato, on his part, thought that it was possible to recognise the truth that is binding for all, and even certainty about the good.

The problem discussed here was to shape ancient thought for centuries: two forms of intellectual authority are confronted here, that of the rhetorician and that of the philosopher. Ultimately, the question was in what form, before which audience and

---

4 Plato, Gorgias, 449c–e, transl. by W. R. M. Lamb.
5 Ibid., 451d.
6 Ibid., 452d/e.
7 Ibid., 454b.
8 Ibid., 466a.

against what background intellectual authority asserted itself. It was also a question of how far a person with intellectual authority had to take the people into consideration or whether he should develop his thoughts independently of them.

Plato liked to refer to negative experiences such as the condemnation of Socrates and refrained from using his intellectual authority at all in the face of a mass such as the Athenian People's Assembly, which lacked the means of knowledge anyway. At best, he sought contact with autocrats such as Dionysius I, a monarch in Sicily, if one is to trust tradition. The contacts of his students were also more strongly oriented towards individual interlocutors, often sole rulers.

Plato refers to another basic problem of intellectual authority, the question of which premises it can develop its line of argument from, and from where certainty can be derived. He developed a complex epistemology, which I cannot discuss here. I only want to highlight one basic concept: in Plato's view knowledge is not empirical, it comes from divine insight. In consequence, it is only accessible to the happy few. This stands in stark contrast to democratic tenets.

In Plato's dialogue Politeia (State), the philosopher imagined an ideal political order in which the philosophers were to be rulers. They were expected to gain all the knowledge necessary to govern the polis. The philosopher king may at first glance appear to be the highest embodiment of intellectual authority – but he is precisely not, since he has such a high level of insight that he can dispense with argumentation. In any case, all this remained far from being realised. There is every reason for the interpretation that Plato did not use this text as a blueprint for a state order anyway. It was a game of thought as to what a good life would require. The knowledgeable exists for Plato in theory, but in the world as it is, one cannot expect him to be heard, certainly not in a democracy, but if at all by individuals who had the skills and resources to assert themselves in his polis.

But most Athenians held on to democracy and tried to change it from within. They remained true to the basic tenets but varied them in different ways. This was not least a matter of redefining rhetoric. Speakers had already been trained by the Sophists. They focused on the argumentation for the probable or plausible. Now rhetoric textbooks were developed which attached importance to the fact that, in addition to probability arguments, there was also evidence, proof. Specialists for certain difficult fields of law, such as inheritance law, came to the fore, even though the idea of a specific legal training was still missing.

Significantly, the speakers who were active after the Peloponnesian War claimed to identify and represent the common interests of the polis, even if this contradicted the current sensitivities of the people: insulting the public was part of the office of the speaker in the People's Assembly. In order to serve the interests of the people, it was necessary for them to distance themselves from the people.

A typical politician of this period was Demosthenes, who has become the epitome of Attic oratory even today. He came from a wealthy family, and although his guardians embezzled some items, his wealth was probably considerable, which he apparently owed to his rhetorical art, with which he was successful in trials against his guardians inter alia. In addition, he earned income with his rhetorical skills, but mainly put it to the service of politics. His appeal to the Athenians to resist Philip II, the ruler of Macedonia and father of Alexander the Great, is famous. But this was to no avail. Athens resisted Philip courageously, but finally suffered a crushing defeat. Later generations celebrated this as an expression of honourable behaviour, but, first of all, it meant horrible bloodshed.

Nevertheless, before and after that, Demosthenes took on the role of a forward-looking admonisher, as in the third speech against the Macedonian King Philip in 341, which refers to a time when the king was very successful. Demosthenes demands that the Athenians should take a stronger stand against him: he criticizes those who study to win your favour rather than to give you the best advice. He himself is different, of course:

> I claim for myself, Athenians, that if I utter some home-truths with freedom, I shall not thereby incur your displeasure. His concluding remarks fit very well: These are my views and these are my proposals, and if they are carried out, I believe that even now we may retrieve our fortunes. If anyone has anything better to propose, let him speak and advise. But whatever you decide, I pray heaven it may be to your advantage.[9]

Demosthenes clearly articulates the claim to be the only one able to recommend the right thing, the only one to possess true foresight. Intellectual authority is openly brought into play here and appeals to the insight of the audience that trusts the good politicians. Through his own higher insight, however, he hopes to attract better citizens who can then act successfully. Demosthenes claimed a role for himself, as Thucydides attributed it to Pericles, but he was much more emotional and moralising, therefore less rational and more dangerous.

Aristotle, a pupil of Plato, takes a completely different position to Plato and Demosthenes. Born in 384, he was, in contrast to Plato and Demosthenes, not a citizen of Athens but was from a city in the northern Aegean, which increasingly came under the influence of the Macedonian Empire. Therefore, he also acted as an advisor to Macedonian rulers, and incidentally also as teacher of Alexander the Great, the world conqueror, whose drunkenness and lack of self-control, however, did not do the philosopher any credit. Aristotle spent many years in Athens and founded a philosophical school, but he remained a foreigner.

Aristotle reflected the role of intellectual authority in the polis thoroughly, above all in his rhetoric, and provided instruments by which representatives of intellectual authority had to be measured.

---

9 Demosthenes, 76; transl. by J. H. Vince.

He also formalised logic, reflecting on the steps of logical argumentation, which served only truth. The same basic ideas also characterise Aristotle's rhetoric, but here it is about the power of persuasion. In contrast to Plato, Aristotle does not design an aloof model but discusses the question of what a speaker can achieve in a city.

Aristotle addresses many aspects in his Treatise on Rhetoric. For my considerations, it is crucial that the philosopher also identifies the limits of the intellectual moment in rhetoric. In contrast to dialectics, which aims for what is right, the goal is to find out what is convincing and plausible. Above all, he distinguishes three conditions of persuasiveness: the overall personality of the speaker, the emotionality of the audience, and finally the lógos, the argument, which is in the end the most important element of persuasiveness. In addition, however, there are also means of persuasion that do not belong to the art of speech in the true sense. These include witness statements or oaths.

The important thing about arguments is that they are based on accepted opinions but are oriented towards what is right, unlike Aristotle's reproach to the Sophists. Furthermore, he reproaches other rhetoricians for relying too much on emotions. In well-ordered cities, he says, the cause must be put centre stage. Suspicion, compassion, anger, and such mental emotions, do not relate to the cause, but are aimed at the judge,[10] the legally untrained common citizen who sat in court. Rhetoric is a matter of conviction, not persuasion, which was central to Gorgias, as characterised in Plato.

Ultimately, Aristotle relies on an elitist habitus of self-control and underlines the right of the wise rhetorician to know better than the others even if he takes the accepted opinions into account. It is, as is often emphasised, a continuation of the Platonic idea that at its core rhetoric should not become arbitrary, but the other perspective should not be overlooked either: like the political rhetoricians of his time, like a Demosthenes, Aristotle holds the accepted views in respect; basically, he trusts the citizens to think correctly. Thus, the orator can make a difference, especially in a well-ordered city – this limitation should be taken seriously. At the same time, the limits of the chances of a man who claims intellectual authority and who is concerned with the cause and not only with success are thus sounded out. Metaphysically founded claims to truth have no place here. But there is the same caveat as with Plato: only if there is an appropriate environment can intellectual authority make a difference.

But Aristotle believes that democracy, along with other political orders, can do this: by attributing the people with the potential for reason and the ability to listen to the advice of the holder of intellectual authority, intellectual authority is incorporated into democracy, even though it can all too easily become a danger to it. In this respect, Aristotle speaks entirely in a democratic tradition, but from an elitist point of view and aware of the danger it poses.

---

10 Aristotle, Rhetoric, 1354a16–18, transl. by H. Leppin.

An elite that constituted itself in the way defined by him was, however, always dependent on acceptance, so that the reproach of others that the speakers only played up to the people was understandable. The tension between the assumed omniscience of all citizens and the existence of an intellectual elite thus remained in existence, although a responsible, rational rhetoric had the potential to build a bridge. The socially disruptive potential of intellectual authority stands on one side, the possibility of fortifying an educational elite, on the other.

So far, we have heard the voices of intellectuals. There was another way to stabilise democracy, namely institutional reforms. The Athenians went down this road; yet we do not know the details of their discussions. Bribery was complicated by machines like the cleroterion, which prevented people from knowing who would sit in court. Athenian courts comprised hundreds of lay people who would give judgement, who would be drawn by lot from an even larger number. Moreover, the Athenians increased the risk of speakers presenting a bad proposal. The speakers could be sued, if they got their way in the popular assembly, for a law against the existing laws. A court of law ruled on this, and the applicant had to expect a severe penalty. Moreover, there were attempts to revert to the old laws and even record them systematically and publicly. In addition, the Athenians invented procedures that made hasty decisions less likely. New laws were systematically checked to see whether they matched the old ones. Democracy, which was in danger of losing its foundation during the Peloponnesian War, sought, like Cleon, security in the laws that were attributed to wiser generations of the old. By considering them as the tenets of an earlier period, they could be varied in such a way as to transform democracy and give it more of a constitutional state. But because in ancient times recourse to the old was held in higher esteem than the claim to achieve something new, it was so important to describe this transformation as a return to the old.

Athenian democracy regained strength even after defeats. It was able to keep most of its citizens loyal and even found imitators elsewhere. The idea of an order based on each citizen being considered competent to participate in it proved remarkably resilient, and it produced great things, even among its opponents.

The fundamental difference between science and politics, that science must orient itself towards the truth, while politics has to balance different interests and make binding decisions, even if these were not optimally justified, was not seen in antiquity. According to the ancient Athenians, it was possible to recognise what was necessary and to make the right decision in a well-ordered state. We know that this is not the case; more recently, the failure of the socialist states, which relied heavily on scientific knowledge, has taught us that in the 20th century. We also know that all scientific knowledge is preliminary and reversible, in contrast to Plato. These differences between antiquity and modernity are fundamental and also legitimise the consideration of the interests of individual groups in modern politics, which ancient observers would have disapproved of.

In consequence, the consideration of history cannot, in any case, provide instructions for concrete actions, but it can remind us of aspects that need to be taken into account when acting, which is all too easily tied to short-term current perspectives. The Athenian example, for instance, may warn against overestimating the moral pathos of a Demosthenes or the self-assurance of a Plato. The efforts of the countless nameless Athenians who worked to reform their political order, in many small steps deserve much more respect. It is here, in the inconspicuous, that democracy lives, involving many with their respective competences, and it is perhaps here that experts are most likely to have their place, not in explaining the world, but in turning small cogs, but most of all, using their professional style of argumentation. The science-based analysis is absolutely necessary for democracies at a time when emotionalisation is so efficacious. It is not only the result of scientific research, but also the habitus of being scientific, which makes experts and scientists indispensable to democracies. As long as democracies are democracies, however, they ultimately depend on the reasoning of the demos, of the people, whom the Athenians trusted and whom I hope we will be able to trust for a long time to come in our world, too.

Further reading

J. Bleicken. Die athenische Demokratie. 4. Aufl., Paderborn 1995.

I. Jordović. Taming Politics. Plato and the Democratic Roots of the Tyrannical Man. (Studies in Ancient Monarchies, Bd. 5.) Stuttgart 2019.

H. Leppin. Paradoxe der Parrhesie, Tübingen (in press).

A. W. Saxonhouse. Free Speech and Democracy in Ancient Athens. Cambridge etc. 2006.

Keynote Speech on 'Tenets, Variations, Transformations'[*]

SANDRA ECKERT[**] / MURIEL MOSER /
CHRISTIAN MÜNCH / HANIEH SAEEDI /
ANDREAS SCHLUNDT / CAMELIA-ELIZA TELTEU[1]

Good afternoon,

It is an honour and great pleasure to give this keynote lecture today, and I thank my colleagues at the Johanna Quandt Young Academy for this opportunity to represent them.

As a first-generation fellow, I would like to begin by sharing some of the experiences we have had in previous years at the Young Academy. Through its events, the Young Academy has been able to drive academic exchange beyond its walls, with the Goethe University, other research partners in the region, and further afield. In October 2019, to give an example, we hosted a two-day workshop with scientists, policymakers and representatives from civil society and business at the Permanent Representation of the State Hessen to the European Union in Brussels. The Academy hopes to resume similar activities in the near future. Such outreach allows us to contribute to addressing the complex problems our society is facing.

Two generations of fellows and members currently form the Academy. We comprise a community of 15 researchers; we come from various disciplines, including biology, history, pharmacy, political science, linguistics, philosophy, physics, media

---

[*] Developed jointly by JQYA members at the Study Days and delivered at the Award Ceremony 2021 by Sandra Eckert on behalf of Academy members.
[**] Sandra Eckert revised the written essay and adapted it into a speech.
1 Further contributors Eva Buddeberg, Philipp Erbentraut, Nadine Flinner, Jasmin Hefendehl, Daniel Merk, Torben Riehl, Peter Smith.

studies, physical geography and medicine; we are affiliated with 14 departments and institutes. Our members hold prestigious grants such as the Emmy Noether and ERC, and are participants in other Young Academies such as The Young Academy of the Berlin-Brandenburg Academy of Sciences and Humanities; the German National Academy of Sciences Leopoldina; the Young Academy of Sciences and Literature Mainz. We fellows are selected based on our outstanding academic achievements, and many of us have an international background. We cover a wide range of research topics, which include: the correlation between cell morphology and function and the underlying molecular features; social memories in the Greek cities during late Hellenistic times; the origins of deep-sea biodiversity; the history of artificial environments and the internet of things; the role of business actors in shaping environmental policy; and the impact of climate change on freshwater resources.

Being part of the Academy gives us extra freedom and independence in conducting our research and addressing these research topics. I have myself had the chance to hold a sabbatical fellowship for one semester, which allowed me to focus entirely on research and publication activities at a crucial stage in my career. Moreover, having access to additional funding is crucial for all of us to implement our research ideas, especially for the natural scientists among us who are often constrained in their activities by the type of expensive equipment they have at their disposal.

As I speak to you today, the Academy is welcoming its third generation of excellent researchers: We welcome eight outstanding academics from sociology, mathematics, history, political science, film and media studies and philosophy. Under normal circumstances, we would meet for a more formal and distinguished ceremony, with music and drinks befitting such a significant occasion. With deep regret, we cannot do so, and all we can share are fond memories of our Opening Ceremonies held in 2018 and the Award Day in 2020, which was held in a format compliant with the special circumstances.

I also have fond memories of the Academy's previous years of 'normal activity' before the pandemic dramatically changed our work routines and daily lives. As of September 2018, we held meetings at the Riedberg Campus on a regular basis. As a social scientist based on the Westend Campus, I must admit that I had reason to visit the scientists' campus for the first time because of the Academy. I have learned a great deal in the discussions with researchers from very different backgrounds, which in the first year of the Academy centred on the theme of 'Nature and Normativity'.

This brings me to the topic of today's lecture, namely the annual themes that guide the Academy's intellectual life and interdisciplinary exchange. Towards the end of our first year in the Academy, we fellows designated the Academy's subsequent theme. This theme was applied throughout 2019 and 2020. The theme we have been discussing in these meetings is structured around three concepts, namely *Tenets, Variations and Transformations*.

These concepts might sound rather abstract to you, and you might have very different ideas about these concepts depending on your own background and experience. This is precisely what instigated the discussions in the Academy when a marine biologist explains her tenets to a philosopher – it is a sense of curiosity, excitement and, at times, even irritation. Let me take you on an intellectual journey to give you a flavour of the Academy's life, and share with you the kind of reflections and thoughts that our group assembled on tenets, variations and transformations.

Tenet is a word that we rarely use in our everyday language. The term originally derives from the Latin word 'tenere', which means 'to hold'. German translations are 'Grundsatz', 'Lehre', or 'Dogma'. A tenet can be regarded as a fundamental principle. Tenets, one could argue, are at the core of a scientific paradigm – if we adhere to the notion suggested by the philosopher of science Thomas Kuhn.

We in the Academy cover a broad range of scientific disciplines, ranging from the natural sciences and linguistics to social and political sciences. The starting point for our discussion was the very basic question: what do we conceive as being a tenet in our discipline? We have come up with manifold answers across disciplines, and I cannot elaborate on all of them due to time constraints. Instead, I will focus on a few.

For the natural scientists among us, there is literally no better example for an overarching tenet than the central dogma of molecular biology: 'DNA makes RNA makes protein'. The protein is generally described as a gene product that defines the phenotype, which are the features of an organism. This is a sentence to be found in textbooks since the late 1970s. Francis Crick initially established the claim, and since then it has been re-postulated in an incomplete way thousands of times. It states that we store and preserve our inheritable genetic information in the form of DNA, while proteins provide the gene products in all possible forms like hair, hormones or enzymes. The messenger RNA mediates the transfer of genetic information from DNA to protein. According to the most basic rules of mathematics and physics – we can call these natural laws – that DNA is transcribed into RNA.

Compared to the 'hard' sciences, our social sciences and humanities tenets often appear more vulnerable and less stable. This also has to do with our objects of scientific enquiry. If we take history as an example, we first have to agree on the *kind of sources we study*. Over the past centuries, the discipline has established an agreement regarding the possibilities and limits of our different sources, texts, materials, etc., in order to achieve reliable research outputs. Secondly, historians have to develop a common understanding – or tenets – as to how they analyse and interpret their objects of studies. Such a common understanding is the result of academic traditions and research experience. Commonly accepted standards allow for comparability. Agreeing on common standards, naturally, is of equal importance in the life sciences. Let me illustrate this with an example from the field of marine biology: global hydrological modelling relies on standardised data collection, which follows the FAIR Principles: Findable, Accessible, Interoperable, and Reusable.

At times, the accumulation of data and their analysis may lead us to question our tenets. Even what we usually consider a natural law is not made for eternity. Let me go back to our DNA example: new findings have shown that the model cannot fully explain the production of gene products from DNA. As a result, the model still holds true in its most principal concept but needs an extension on other aspects such as temporal and spatial variation; it also requires an extension on the tenet because proteins are not the only molecules that define the phenotype. How do we best respond to such new findings? Are we better off putting certain tenets to one side to allow novel ones to inform our concepts?

In the humanities and social sciences, we are used to the idea that our tenets are not matters of fact but rather represent the current state of knowledge in our discipline and that different tenets may even coexist in line with distinct theoretical approaches. But also in the natural sciences, as our example has illustrated, we need to put our hypotheses to empirical scrutiny. In 'normal' times – to use the terms of Thomas Kuhn, who has discussed the history of science mainly referring to examples of the natural sciences – this might mean that we adapt our tenets, but we do not need to reject them. In times of 'scientific revolutions', by contrast, we have to fundamentally change our tenets as we move to a different paradigm of doing science. Galileo Galilei's achievements are probably the most eminent example of such a scientific revolution.

The situation becomes even more complicated once we engage in interdisciplinary exchange. While we might find it hard to share common tenets across disciplines, this conversation can help us sharpen our understanding of other disciplines' assumptions and basic principles – or even to become more aware of our own ones. Such awareness can be crucial to understand the limitations of our field of research. This exchange might, under certain conditions, even allow us to come up with tenets that are informed by different disciplines. If this is the case, we transgress interdisciplinary practice and engage in truly trans-disciplinary research. Tenets, then, are difficult but necessary concepts for scientific research and debate. They are necessary because, to arrive at meaningful results, scientists need to formulate abstract, simplifying, and often simplistic tenets. The process of reduction and standardisation is a necessary precondition for categorisation, scientific inquiry and theorisation.

Variation in nature and human behaviour represents perhaps the greatest challenge for scientific research. At the same time, tenets must be grounded in reality; and *reality* – life, nature and society – *is* full of variations. These variations result from evolution, chance and change, and are a necessary requirement for survival, even in the smallest units of our existence, our DNA. As a result, scientific research needs to take variations into consideration.

To begin with, when defining their object of study, scientists need to develop models of variation, that is, classifications and typologies. Indeed, nature cannot be studied without reference to its variations. Let me give an example from hydrology: global water models try to simulate the variations in the terrestrial water cycle on the global scale while excluding the ocean component and the water quality. History, society, and politics cannot be analysed in any meaningful way without attention to variation. Historians study humans, who are erratic, impulsive and irrational objects of analysis. Variation is thus the norm in historical sources. This, together with the contingency factor of the surviving material, means that while there are certain patterns or rhetorical structures that can be observed – such as in literary words, inscriptions or coins – no source is like another; every source is *unique*. When studying political systems, to give another example, we usually classify them according to fundamental features such as democracy versus authoritarianism. Moreover, we differentiate democratic regime types further. Often times we do so based on dichotomous typologies, such as the one contrasting parliamentarian versus presidential systems, which focuses on the interaction between legislature and executive.

These examples show how in our research, we try to make sense of variation. Typically, we come up with a more abstract model, typology or classification, which then serves as an analytical tool that allows us to deal with the kind of variation we observe. Biology can help us elaborate on this: in order to analyse the behaviour of cells, it is necessary to define their state of being in the first place. One such possibility is the classification of a specific cell type based on morphological criteria or the expression of a certain protein. Through the advancement of methodology, it is now possible to track the morphology and the genetic profile of the cells in question, longitudinally and with high resolution. The use of these advanced methods has shown that many cell types display variations of the described initially cellular morphology, as well as transformations of their genetic signature. These transformations occur in relation to variables such as time or a damaging event (like UV-light damage on DNA). Based on the occurrence of such transient or permanent variations within the cellular morphology and function, one has to ask the question of whether the original classification will still hold true for all identified variations of the cell type. Hence, many scientists now rely on the creation of subclasses to accommodate newly discovered variations of an original cellular phenotype. Even though the initial classification of each specific cell type is necessary to form working hypotheses, a new school of thought advocates a more fluent classification in which cells can transiently move in and out of states of activation.

Variation thus highlights the limitations of attempting to classify observed reality. Our schemes of classifying and identifying types impose constraints on research already at the analytical level. Moreover, how we analyse our object of study will differ depending on the schools of thought or theoretical approach we adhere to. While the object studied – be it a historical figure, language or a cell type – remains the same, different schools of thought or theories influence how the object is being studied. For example, in global water modelling, models and the model equations used to describe various processes take very different forms. Such variety has hampered a comprehensive understanding of how models operate, why they differ in their simulations, and how they can be improved. In the humanities, such as history, the increasing salience of notions such as gender or space re-orient analytical approaches. Ultimately, this change of view of the observer is conducive to different interpretations of a historical text. These examples illustrate that the same subject matter can be analysed using different approaches and methods, producing a variation in scientific models and the kind of findings these generate.

While such variety can be problematic in making exchange and comparability difficult, it is also a benefit. Scientific progress is not possible without variation. Variation is *a*, if not *the*, necessary precondition for the advancement of knowledge. The more research approaches, models, teams and groups there are, the more potentially useful insights can be gained from them.

An appropriate example of this is the current global search for vaccines against Covid-19. Here, coexisting but also collaborative research efforts have resulted in the swift development of a number of new different vaccines. Most importantly, varying approaches will be crucial in coping with new coronavirus variants. This last example, which links directly to the most pressing challenge we face at this very moment, the global Covid-19 pandemic, brings me to the last key concept of this talk: *transformation*.

In our view, the type of transformation which is fuelled by an institution such as the Young Academy is twofold: it is about the transformation of us, the fellows and members of the Academy; and it is about the transformation of society, to which *we* as researchers can make a significant contribution. Let me first explain how *we* are being transformed through the Young Academy. As mentioned earlier, there is a need for communication, explanation and translation in an interdisciplinary context such as the Academy. This is very much a matter of finding a common language. This effort begins at the fundamental level of assumptions and ontology (tenets), covers our units of analysis (variations), and subsequently requires a transformation of ourselves. We need to be able to communicate in a way that is intelligible to scientists from other disciplines and the wider society; in a way that is 'theory-free', abstracting from specific theoretical or dogmatic paradigms. Going beyond a narrow disciplinary perspective by reflecting on our (disciplinary) tenets and variations from a meta-level can allow us to engage in a truly *trans*disciplinary discourse and research practice. From a philosophical standpoint, this presumes that communication between sub-systems – in this case, the highly specialised disciplines in the world of academia – *is* possible. As a Frankfurt-based Young Academy, we adhere to this Habermasian dictum.

Only when we can speak to each other and communicate effectively will we draw on the benefits of this interdisciplinary endeavour. And these benefits go beyond the scientific and research community; these are rewarding for society: by engaging in an exchange with other disciplines, we can raise our awareness of societal problems or even learn of their existence altogether.

Let me give an example: when a political scientist from the Westend Campus interested in sustainability issues encounters a scientist from the Senckenberg Research

Institute who is conducting research on deep-sea biogeography, the synergies may be surprising and groundbreaking. The social scientist, for obvious reasons, will not be aware of the kind of sophisticated scientific data and the problems encountered in data collection and analysis of marine biodiversity. However, understanding the key issues at stake will help her a great deal in perceiving and understanding what the policy problems might be and getting a sense of what evidence-based policymaking could look like in this specific context. Likewise, the scientist might get a sense of *why* scientific findings do not directly translate into policy outputs and that scientific evidence is likely to be used strategically in the political power play.

Beyond interdisciplinary exchange, a truly transdisciplinary perspective might be even more rewarding where transgressing disciplinary boundaries allows us to generate new types of knowledge. Let us remain with the example of biodiversity: our marine scientist from Senckenberg might change her problem perception altogether when cooperating with our social scientist, and vice versa. In short: an integrative view on policy problems, such as protecting biodiversity, would not only mean to aggregate the distinct problem perceptions of separate disciplines but to overcome them and take a more encompassing view.

Ultimately, such a transformation of ourselves as fellows and members of the Academy is therefore conducive to a transformation in our research practice, and by that means, a significant contribution can be made to transformative processes at the societal level. With better and more encompassing evidence about *what* these problems are, we can make a difference and will also be able to communicate the societal benefits of our research to a wider audience effectively.

And this is where the circle closes – and where I come close to the end of this lecture. One thing we have learned in the Young Academy is to speak in a comprehensible way about our highly specialised everyday research practice. The Young Academy's exchange has helped us take a position in public discourse, drawing on multiple perspectives and not being bound by the narrow confines of a single discipline. This enables us to confront new and unexpected perspectives, so that not *individually* but rather as a *group* we can contribute to and even spark public discourse, and all this thanks to the Academy.

I want to conclude by saying that we fellows and members are deeply grateful to the Johanna Quandt Foundation for providing us with this unique opportunity to establish links across departments, universities and research institutions and to engage in a fruitful interdisciplinary exchange. We are thrilled to see the Johanna Quandt community growing and express our warmest welcome to the new fellows, members, distinguished senior scientists and ambassadors. The second-generation fellows and members very much look forward to continuing the debate in the next academic year on a new annual topic, which has yet to be defined.

On behalf of the first generation of fellows, I would like to express our gratitude and assure you that we will be delighted to, and committed to, continue playing an active part in the Academy's life as alumni. Thank you very much.

Grundsätze, Variationen, Transformationen
Überlegungen zur Arbeit und Rolle der
Johanna Quandt Young Academy in Wissenschaft
und Gesellschaft
Mit Vorwort von Herbert Zimmermann

MURIEL MOSER* / SANDRA ECKERT /
CHRISTIAN MÜNCH / HANIEH SAEEDI /
ANDREAS SCHLUNDT / CAMELIA-ELIZA TELTEU**

Vorwort

Volluniversitäten wie die Goethe-Universität zeichnen sich durch einen breiten Fächerkanon aus, zu dem unter anderem die Geisteswissenschaften, Wirtschafts- und Sozialwissenschaften, Jura, Mathematik, Naturwissenschaften und Medizin gehören, teilweise auch die Ingenieurwissenschaften. Diese werden ergänzt und erweitert durch außeruniversitäre Forschungseinrichtungen. Es ist ein Merkmal deutscher Universitäten, dass all diese Fächer unter dem einen Dach der Alma Mater beforscht, gelehrt und verwaltet werden. Was einmal klein angefangen hat, hat sich zur Massenuniversität ausgewachsen, mit mehreren zehntausenden Studierenden und hunderten Professorinnen und Professoren. Die zunehmende Anzahl an Fächern und Hochschullehrerinnen und Hochschullehrern erschwert den gedanklichen Austausch und damit die Entwicklung kreativer Ideen, die über die gewohnten Bahnen des eigenen Forschungsansatzes hinausweisen. Die fachliche führt zur sprachlichen Spezialisierung, die sich zu einer babylonischen Sprachverwirrung auszuwachsen droht, inmitten derer

---

* Leitung.
** Weitere Autor*innen: Eva Buddeberg, Philipp Erbentraut, Nadine Flinner, Jasmin Hefendehl, Daniel Merk, Torben Riehl, Peter W. Smith.

sich Wissenschaftlerinnen und Wissenschaftler am Ende nicht mehr verstehen. Hinzu kommt häufig der Verlust der räumlichen Einheit, die Verlagerung der einzelnen Einrichtungen in unterschiedliche Teile einer Großstadt.

Die Universitäten ergreifen Gegenmaßnahmen, um den fachübergreifenden Austausch zu forcieren. Sie richten fachübergreifende Studiengänge ein, bilden fachübergreifende Zentren und Forschungsvorhaben, die meist aber doch nur wenige Fachgebiete umfassen. Gibt es auch Räume für den nicht an spezifische Forschungsvorhaben gekoppelten breiten transdisziplinären Diskurs, insbesondere für Nachwuchswissenschaftlerinnen und Nachwuchswissenschaftler?

Die Johanna Quandt Young Academy at Goethe bietet diesen Raum einer ausgewählten Gruppe exzellenter junger Wissenschaftlerinnen und Wissenschaftler aus ganz unterschiedlichen Fächern, die bereits ihren eigenen Weg gefunden haben und am Beginn einer wissenschaftlichen Karriere stehen. Die Treffen und die Bearbeitung selbstbestimmter Themen stimulieren den Gedankenaustausch zwischen Disziplinen, die sonst kaum oder nie interagieren würden. Dieser Gedankenaustausch ist von grundsätzlicher Bedeutung, sollen sich unterschiedliche Wissenschaftsbereiche doch gegenseitig befruchten und neue, vielleicht zuvor nicht gedachte Ideen und Konzepte hervorbringen. Das ist das eine. Das andere ist die Verantwortung, die Wissenschaftlerinnen und Wissenschaftler gegenüber ihrer eigenen Gesellschaft und auch im globalen Kontext haben. Zu den Themen, die gegenwärtig von besonderer Aktualität sind, gehören etwa der Klimawandel, die Biodiversität, das Gesundheitswesen einschließlich der SARS-CoV-2-Pandemie oder die Menschenrechte. Hier muss die kommende Generation wissenschaftlicher Führungskräfte ihren Beitrag leisten.

Das einführende Kapitel des Beitrages reflektiert das letzte Jahresthema der Akademie: Grundsätze, Variationen und Transformationen (Tenets, Variations and Transformations). Es untersucht in einem fächerübergreifenden Ansatz Grundsatzfragen zum transdiziplinären Diskurs und der daraus resultierenden Wissenschaftsentwicklung. Es zeigt, wie die Weiterentwicklung der Forschungsansätze einzelner Wissenschaftlerinnen und Wissenschaftler schließlich zu einem Transformationsprozess auf persönlicher und gesellschaftlicher Ebene führen kann. Dafür muss um einen Konsens der Begrifflichkeiten gerungen werden, der es erst ermöglicht, über alle wissenschaftlichen Disziplinen hinweg in Dialog zu treten. Am Beispiel des Begriffs Tenet bzw. Grundsatz wird dies verdeutlicht. Welche Bedeutung haben in den einzelnen Fächern Grundsätze, Modelle und Dogmen? Wie wichtig sind Variationen (variations) dieser Konzepte für die Weiterentwicklung der Wissenschaft und wie werden diese in den unterschiedlichen Fächern implementiert? Erst die Auseinandersetzung mit diesen Fragen führt zu einem neuen wechselseitigen Verständnis (transformations), das es den Beteiligten ermöglicht, einen fruchtbaren transdisziplinären Diskurs und eine transdisziplinäre Forschungspraxis zu etablieren, einschließlich der Transformation im öffentlichen Raum. Dies wird mit eindrucksvollen Beispielen belegt.

Es freut mich besonders, dass die Young Academy und die Wissenschaftliche Gesellschaft enger zusammengerückt sind. Die Wissenschaftliche Gesellschaft ist offen für alle akademischen Fächer und pflegt ähnlich wie die Young Academy den transdisziplinären Diskurs. Der Dialog zwischen exzellenten jungen Führungskräften und etablierten Vertreterinnen und Vertretern ihres Fachs wird für beide Seiten fruchtbar sein.

Ich wünsche der Young Academy ein langes Leben, den jungen Wissenschaftlerinnen und Wissenschaftlern den besten Erfolg mit ihren Arbeiten und ihrer weiteren Laufbahn und schließlich den Mut, etablierte Konzepte auf den Prüfstand zu stellen.

Professor Dr. Herbert Zimmermann
Präsident der Wissenschaftlichen Gesellschaft an der Johann Wolfgang Goethe-Universität

Einleitung

In diesem Beitrag soll die Arbeit der Johanna Quandt Young Academy (JQYA), fortan Young Academy, vorgestellt und erklärt werden, warum sie für uns, die Fellows der Akademie, für die Universität und die Wissenschaftsgemeinschaft im Allgemeinen von Bedeutung ist. Die Young Academy ist eine von der Johanna Quandt Stiftung geförderte und an der Goethe-Universität Frankfurt (GU) angesiedelte Einrichtung. Sie ist offen für erfahrene Nachwuchswissenschaftler*innen der GU und von Partnerinstitutionen wie den Max-Planck-Instituten, der Technischen Universität (TU) Darmstadt, der Universität Mainz und des Senckenberg Instituts.

Der Nutzen der JQYA ist ein doppelter. Erstens unterstützt sie uns, ihre Mitglieder, dabei, unsere eigene Forschung besser zu verstehen; und zweitens ermöglicht sie uns eine Auseinandersetzung mit den einzelnen Disziplinen nicht nur auf akademischer, sondern auch auf gesellschaftlicher Ebene. Dies wird durch eine interdisziplinäre Reflexion über Forschung und die ihr zugrundeliegenden Prinzipien möglich gemacht. Diese folgt dem Jahresthema der Akademie, das in diesem Jahr (2020/21) „Grundsätze, Variationen und Transformationen" („Tenets, Variations and Transformations") lautete.

Anhand dieses Jahresthemas soll im Folgenden die Arbeit innerhalb der Akademie vorgestellt werden. In den ersten beiden Abschnitten stellen wir vor, wie wir gemeinsam über Grundprinzipien nachdenken – in diesem Jahr über Grundsätze und deren Variationen in Forschung und Wissenschaft. Dieses Thema wirft gerade im interdisziplinären Austausch zahlreiche epistemologische Fragen auf. In Abschnitt 3 legen wir dann dar, wie diese transformative Debatte uns als einzelne Forscher*innen verändert hat. Zuletzt wird Teil 4 darstellen, wie diese Veränderung für Frankfurt, die Wissenschaft und die Gesellschaft im Allgemeinen von Nutzen ist.

1.   Grundsätze („Tenets")

„Tenet" ist ein Wort, das wir in unserer Umgangs-, Alltags- oder auch Wissenschaftssprache selten verwenden. Der Begriff leitet sich ursprünglich vom lateinischen Wort
„tenere" ab und bedeutet „halten". Wenn wir den Begriff heute in eine Google-Suche
eingeben, erhalten wir die deutschen Übersetzungen „Grundsatz", „Lehre" oder „Dogma". Ein Grundsatz kann einerseits als ein fundamentales Prinzip betrachtet werden.
Andererseits kann der Begriff Grundsatz auch ein Werkzeug (Datensatz) bezeichnen,
das zur Gewinnung neuer Erkenntnisse oder zur Validierung bestehenden Wissens
verwendet wird.

Diese Mehrdeutigkeit des Begriffs „Tenet" kann die Kommunikation über Disziplinen hinweg schwierig machen. Uns, den Mitgliedern der Young Academy, ermöglicht sie aber auch, bereichernde Diskussionen zu führen und neue Blickwinkel auf
verschiedene Fragestellungen zu gewinnen. Unsere Mitglieder, Fellows und Members
decken ein breites Spektrum an wissenschaftlichen Disziplinen ab, das z. B. von den
Naturwissenschaften über die Sprachwissenschaften bis hin zu den Sozial- und Politik-
wissenschaften reicht. Dies erfordert die Definition eines Pools von gemeinsamen Begriffen, die für interdisziplinäre Diskussionen genutzt werden können. Eine gemeinsame Sprache erleichtert die Kommunikation, erfordert aber die korrekte Übersetzung
der relevanten Begriffe. Vor einer solchen Herausforderung stehen Politiker*innen in
der internationalen Politik, aber auch Wissenschaftler*innen, und sie beginnt natürlich mit der ganz grundsätzlichen Frage: Was verstehen wir als Grundsatz, als logischen Ausgangspunkt für jede einzelne, immanente wissenschaftliche Struktur? Dies
soll hier anhand einiger repräsentativer Spezifika rund um Grundsätze innerhalb unserer Disziplinen illustriert werden. Zudem wird deren unterschiedliche und mehrdeuti-
ge Verwendung skizziert.

Auf einer basalen Ebene sind Grundsätze eine wesentliche Voraussetzung für wissenschaftliches Arbeiten. Sie kommen in zwei Formen vor: Zum einen wurde in den
vergangenen Jahrhunderten eine Reihe von Grundsätzen über die Möglichkeiten und
Grenzen unserer verschiedenen Quellen, Texte, Zellen, Materialien etc. aufgestellt
(nennen wir sie Quellenlehren). Diese Grundsätze sind nicht vage, sondern zwangsläufig unbeweglich, um zuverlässige Forschungsergebnisse zu garantieren. Es gibt jedoch die Möglichkeit, verschiedene Werkzeuge oder Ansätze zu kombinieren, um Einschränkungen zu reduzieren. Z. B. können Historiker*innen literarische Quellen neben
archäologischem und epigraphischem Material verwenden oder mit Modellen aus anderen Disziplinen arbeiten, um aus bekanntem Material neue Ergebnisse zu gewinnen.

Zweitens gibt es Lehrmeinungen bezüglich der Interpretation unserer Studienob-
jekte (nennen wir sie interpretative oder analytische Lehrmeinungen). Einige davon
sind das Ergebnis von akademischen Traditionen und Erfahrungen. Die Grundsätze
dieser Kategorie könnten als Werkzeuge bezeichnet werden. Als solche schaffen sie
einen allgemeingültigen Standard. Sie ermöglichen Vergleichbarkeit, während sie

unterschiedliche Sichtweisen oder Haltungen im Allgemeinen nicht beeinträchtigen. Dennoch folgt das Werkzeug hier dem Prinzip. Daher werden z. B. bei der globalen hydrologischen Modellierung und der marinen Biodiversität einige zentrale Grundsätze aufgestellt, die die wissenschaftliche Forschung erleichtern. Diese sind 1) standardisierte Datenerfassung; 2) Adressierung globaler Berichtsanforderungen; 3) Berücksichtigung nationaler und globaler Politik, Berichtsanforderungen und Vorschriften; 4) Daten FAIR machen: Anpassung der Daten an die FAIR-Prinzipien (Findable, Accessible, Interoperable, and Reusable); 5) zugängliche analytische Algorithmen, Werkzeuge und Arbeitsabläufe; 6) Bereitstellung politikfähiger Vorschläge; 7) Zusammenarbeit zwischen Wissenschaft, Industrie und Öffentlichkeit.

Manchmal werden Grundsätze zu „Dogmen" oder Glaubenssätzen. Ein Beispiel kann dies verdeutlichen. Für die Naturwissenschaftler*innen unter uns gibt es kein besseres Beispiel für einen übergreifenden Glaubenssatz als das zentrale Dogma der Molekularbiologie: „DNA macht RNA macht Protein" ist ein Satz, der seit den späten 1970er Jahren in Lehrbüchern zu finden ist. Er wurde ursprünglich von Francis Crick formuliert und ist seither in unvollständiger Form tausendfach neu postuliert worden. Er besagt, dass wir unsere vererbbare genetische Information in Form von DNA speichern und bewahren, während die Genprodukte, also Proteine, die Ausprägung eines Organismus direkt oder durch Steuerung metabolischer, molekularer etc. Prozesse bestimmen. Die Proteine werden in allen möglichen Formen wie Haaren, Hormonen oder Enzymen, aber auch als Genexpression-regulierende Faktoren hergestellt. Die Übertragung der genetischen Information von der DNA auf das Protein erfolgt durch die sogenannte Boten-RNA, auch mRNA genannt. Dieses Grundprinzip gilt für alle Lebewesen. Dies veranlasste Crick und uns, dafür den Begriff „Dogma" zu verwenden. Wir gingen zwar nicht soweit, es als Naturgesetz zu bezeichnen. Und doch sind es Naturgesetze, die diesem Dogma zugrunde liegen. Z. B. sind es die grundlegendsten Regeln der Mathematik und Physik, die es ermöglichen, dass DNA in RNA umgeschrieben werden kann. Andere Beispiele sind Denkschulen in Geschichte, Politikwissenschaft oder Philosophie. Hier kann es eine Kodifizierung dessen geben, was „gute Geschichte" oder „gute Philosophie" ist: die Geschichte des Staates, der Institutionen, der Monarchen (mächtige Subjekte) ODER der Gesellschaft, der Frauen, der Minderheiten, der Kinder (d. h. der nicht-mächtigen Subjekte), oder die Philosophie der Natur gegenüber der Philosophie des Geistes.

Dennoch sind Grundsätze, Modelle oder Dogmen nicht für die Ewigkeit gemacht. Neue Erkenntnisse haben gezeigt, dass ein Teil dieser Grundsätze falsch ist. Wir haben inzwischen gelernt, dass die Natur selbst z. B. den Crick-Grundsatz in Frage stellt und die Genexpression, also die Herstellung von Genprodukten aus der DNA, nicht mehr vollständig durch das Modell zu erklären ist, während das Modell an sich grundsätzlich immer noch gilt. Wir sehen jedoch, dass die Genexpression zeitlich und räumlich stark gesteuert werden muss und dass gerade diese regulatorischen Aspekte nicht von dem traditionellen Grundsatz erfasst werden. Außerdem gibt es eindeutige Hinweise auf Nicht-Protein-Genprodukte, die für den Phänotyp des Organismus ebenso relevant sind.

Diese Variation in der Natur – ein Thema, auf das später noch näher einzugehen ist – zwingt uns, über unsere Grundsätze, Modelle und Dogmen nachzudenken. Sollen wir bestimmte Grundsätze besser beiseitelassen und neue in unsere Konzepte aufnehmen? Sollen wir die Grundsätze im Laufe der Zeit variieren? Manche Disziplinen, insbesondere die Geisteswissenschaften, sind sich oft bewusst, dass ihre Grundsätze keine Tatsachen sind, sondern dass sie den aktuellen Wissensstand einer Disziplin widerspiegeln. Daher können und müssen Grundsätze an neue Erkenntnisse angepasst werden. Aber ist es dann sinnvoll, von „Grundsätzen" zu sprechen? Widerspricht diese Skepsis nicht der Idee von Grundsätzen? Darauf gibt es keine richtige oder falsche Antwort, zumindest nicht für eine oder einen Naturwissenschaftler*in, und möglicherweise auch für keine Wissenschaftler*innen überhaupt.

Noch interessanter wird die Situation im interdisziplinären Austausch. Denn abgesehen von unserer Unsicherheit bei der Verwendung von „Dogma" als Äquivalent zu „Grundsatz", wird die Ambiguität des Wortes „Dogma" sofort deutlich, wenn man die Perspektiven verschiedener Disziplinen auf das Wort vergleicht. Theolog*innen mögen ihre Stimme erheben und behaupten, „Dogma" sei ein unumstößlicher Glaube, z. B. an Gott als nicht verhandelbare Tatsache, die nicht von Menschen beeinflusst und gestaltet werden darf. In der Politikwissenschaft wiederum wird unter „Dogma" eher eine Behauptung oder ein Grundsatz verstanden, der von einzelnen Personen aufgestellt wird und somit nicht unbedingt universelle Gültigkeit besitzt. Diese Überlegungen zeigen, dass wissenschaftliche Erkenntnisse einfach, klar, selbsterklärend und einheitlich kommuniziert werden müssen. Dies kann durch die Entwicklung einer einheitlichen Schreibweise der Modellgleichungen oder durch die Verwendung des gleichen Symbols für den gleichen Vorgang geschehen. Eine solche Beschreibung der Grundsätze wäre für alle Seiten von Vorteil. Ist dies möglich? Einen maßgeblichen Grundsatz gibt es bereits über alle wissenschaftlichen Disziplinen hinweg: die Notwendigkeit einer guten wissenschaftlichen Praxis. Diese besteht in einer Reihe von selbst auferlegten Regeln, die für Wissenschaftler*innen unabhängig von ihrem Betätigungsfeld gelten. Das ist keine Kleinigkeit: Es ist dieses Streben nach gemeinsamen Prinzipien, das eine Symbiose innerhalb und zwischen den Disziplinen ermöglicht, das Anknüpfungspunkte und die notwendige Überschneidung für die Bereitschaft zum Austausch miteinander schafft.

2.   Variationen

Grundsätze sind also ein schwieriges, aber notwendiges Konzept für die wissenschaftliche Forschung und Debatte. Sie sind notwendig, weil Wissenschaftler*innen, um zu sinnvollen Ergebnissen zu kommen, Modelle oder Grundsätze formulieren müssen, die abstrakt, vereinfachend und oft simplifizierend sind. Es handelt sich also nicht um verlässliche Repräsentationen der Realität, auch wenn dieser Prozess der Reduktion und Standardisierung eine notwendige Voraussetzung für Kategorisierung, wissenschaftliche Untersuchung und Theoretisierung ist. Folglich hat dieser Ansatz seine Grenzen. Denn seine grundlegenden Modelle und Grundsätze müssen in der Realität verankert sein und die Realität, das Leben und die Natur sind voller Variationen. Diese Variationen sind ein Ergebnis der Evolution, des Zufalls und der Veränderung. Sie sind eine notwendige Voraussetzung für das Überleben, selbst in den kleinsten Einheiten unserer Existenz, unserer DNA. Variation in der Natur und im menschlichen Verhalten stellt somit die vielleicht größte Herausforderung für die wissenschaftliche Forschung dar. Daraus ergibt sich die Notwendigkeit, Variationen in der wissenschaftlichen Forschung zu berücksichtigen und zu reflektieren, auf welche Weise Grundsätze und Variationen in unseren wissenschaftlichen Bemühungen zusammenhängen.

Unsere Diskussion ergab mindestens vier Ebenen ihrer Interaktion. Erstens gibt es Modelle der Variation; zweitens gibt es ein Spannungsverhältnis zwischen Modellen und Variationen; drittens gibt es die Variation wissenschaftlicher Modelle; und schließlich ist die Variation oder die interdisziplinäre Debatte ein Mittel, um sich mit den Herausforderungen der Variation fruchtbar auseinanderzusetzen.

Zu Beginn müssen Wissenschaftler*innen bei der Definition ihres Studienobjekts Modelle der Variation entwickeln. In der Tat kann die Natur ohne Bezug auf ihre Variationen nicht effizient untersucht werden. So wie beispielsweise globale Wassermodelle versuchen, die Variationen des terrestrischen Wasserkreislaufs auf der globalen Skala zu simulieren (wobei die Ozeankomponente und die Wasserqualität ausgeklammert werden), erfordert ein grundlegendes Verständnis von Biodiversitätsmustern, Konnektivität, Naturgeschichte und Evolution die Untersuchung der Zusammenhänge zwischen Variationen von Lebensräumen einerseits und deren Co-Abhängigkeit von der Variation der Biodiversität andererseits. Auch das menschliche Verhalten kann ohne die Berücksichtigung der Variation nicht sinnvoll analysiert werden.

Wie kann mit dieser Sachlage umgegangen werden? Erstens können Wissenschaftler*innen diese Variation als analytisches Werkzeug nutzen. Zweitens können sie sich dafür entscheiden, die Variation zum Gegenstand der Untersuchung zu machen. Drittens können sie auch genügend Material zusammentragen, um die zugrundeliegenden Gemeinsamkeiten sinnvoll zu kommentieren, oder viertens können sie sich auf eine Fallstudie konzentrieren – mit allen Einschränkungen für die Generalisierbarkeit. An zwei Beispielen aus den Geisteswissenschaften lässt sich dies verdeutlichen. Historiker*innen erforschen Menschen – unberechenbare, impulsive und irrationale Analyseobjekte. Variation ist daher die Norm in historischen Quellen. Zusammen mit dem Faktor der Kontingenz des überlieferten Materials bedeutet dies, dass es zwar bestimmte Muster, rhetorische Strukturen, literarische Modelle oder Ähnliches gibt, die beobachtet werden können, wie z. B. in literarischer Sprache, Inschriften oder Münzen; jedoch ist keine Quelle wie die andere, jede Quelle ist einzigartig. Variation ist auch ein grundlegendes Prinzip in der Erforschung des Zeitgeschehens, etwa in der vergleichenden politikwissenschaftlichen Forschung. Der Vergleich und damit die Variation kann sich auf bestimmte Analyseeinheiten wie Länder, aber auch auf Politikbereiche oder auf Akteurstypen beziehen. In der vergleichenden politischen Ökonomie – um ein Beispiel zu nennen – widmet sich ein Forschungszweig den Spielarten des Kapitalismus.

Allerdings wird Variation oft nicht berücksichtigt, wenn es darum geht, den Gegenstand der wissenschaftlichen Untersuchung festzulegen oder die Methodik, mit der dieser untersucht werden soll. Dann stellen Variationen ein Problem dar, da sie offenbaren, dass diese Modelle die Realität nicht vollständig erfassen. Es besteht dann ein Spannungsverhältnis zwischen Modellen und Variationen. Dies ist z. B. in der biologischen Forschung der Fall, wo es notwendig ist, einen Ist-Zustand für Zellen zu definieren, um deren Verhalten analysieren zu können. Ein Beispiel dafür ist die Klassifizierung eines bestimmten Zelltyps anhand von morphologischen Kriterien oder der Expression eines bestimmten Proteins. Durch die Weiterentwicklung der Methodik ist es heute möglich, die Morphologie und das genetische Profil der betreffenden Zellen in Längsrichtung und mit hoher Auflösung zu verfolgen. Der Einsatz dieser fortschrittlichen Methoden hat gezeigt, dass viele Zelltypen sowohl Variationen der ursprünglich beschriebenen Zellmorphologie als auch Veränderungen ihrer genetischen Signatur aufweisen, die in Abhängigkeit von Variablen wie Zeit oder einem Schadensereignis auftreten. Aufgrund des Auftretens solcher transienten oder permanenten Variationen innerhalb der zellulären Morphologie und Funktion muss man sich die Frage stellen, ob die ursprüngliche Klassifikation für alle identifizierten Variationen des Zelltyps noch zutreffend ist. Daher behelfen sich viele Wissenschaftler*innen mit der Schaffung von Unterklassen, um neu entdeckte Variationen eines ursprünglichen zellulären Phänotyps unterzubringen. Obwohl die anfängliche Klassifizierung jedes spezifischen Zelltyps notwendig ist, um Arbeitshypothesen bilden zu können, zeichnet sich eine neue Denkschule ab, in der eine fließendere Klassifizierung vorgeschlagen wird, in der sich Zellen vorübergehend in Aktivierungszustände hinein und heraus bewegen können.

Variationen zeigen somit die Grenzen des Versuchs auf, die Realität zu klassifizieren. Sie verdeutlichen, wie Klassifizierung, Vorauswahl und andere Vorannahmen die Forschung einschränken, auch auf der analytischen Ebene. Z. B. haben Denkschulen einen großen Einfluss auf die Forschung. Während das Objekt, das untersucht wird – sei es eine historische Figur, eine Sprache oder ein Zelltyp –, dasselbe bleibt, beeinflussen verschiedene Denkschulen oder Dogmen die Art und Weise, wie das Objekt untersucht wird. Sie geben somit die Richtung der wissenschaftlichen Untersuchung und Beschreibung vor. In der Wissenschaft haben auch neue Technologien oft einen großen Einfluss auf die Art und Weise, wie die Objekte untersucht werden. In den Geisteswissenschaften wie in der Geschichte oder der Politikwissenschaft beeinflusst wiederum die zunehmende Bedeutung von Begriffen wie Geschlecht oder Raum, wie

Wissenschaftler*innen ihr Material betrachten. Sie führen zu anderen Interpretationen eines historischen Textes und bieten durch den veränderten Blickwinkel der Betrachter*innen andere Möglichkeiten der Analyse.

Wie das Leben voller Variationen ist, so ist es auch die Wissenschaft. Ein und derselbe Gegenstand kann mit zahlreichen unterschiedlichen Ansätzen und Methoden analysiert werden, was zu einer Variation von wissenschaftlichen Modellen führt. Die Variation von Modellen kann ein Problem sein: Zu viele Modelle desselben Untersuchungsgegenstandes behindern den wissenschaftlichen Austausch, die Zusammenarbeit und synergetische wissenschaftliche Entwicklungen. Dies ist z. B. bei der globalen Wassermodellierung zu erkennen, wo Modelle und Modellgleichungen, die zur Beschreibung verschiedener Prozesse verwendet werden, unterschiedliche Formen annehmen. Diese Faktoren behindern eine umfassende Verständigung darüber, wie Modelle funktionieren, warum sie sich in ihren Simulationen unterscheiden, wie sie verbessert werden können. Nicht zuletzt erschweren sie den wissenschaftlichen Austausch, der eine Voraussetzung für produktive akademische Forschung ist.

Variationen sind aber auch ein Vorteil, denn ohne Variation ist wissenschaftlicher Fortschritt nicht möglich. In der Tat ist eine Vielfalt wissenschaftlicher Untersuchungen und Forschungen eine, wenn nicht DIE notwendige Voraussetzung für den Fortschritt des Wissens. Je mehr Forschungsansätze, Modelle, Teams, Gruppen usw. es gibt, desto mehr potenziell nützliche Erkenntnisse können daraus gewonnen werden. Ein passendes Beispiel dafür ist die aktuelle weltweite Suche nach Impfstoffen gegen Covid-19. Hier hat die Variation zu einer Zusammenarbeit zwischen den Forschungseinrichtungen und zu einer Koexistenz mehrerer Impfstoffe geführt. Und Variation wird notwendig sein, um neuen Variationen des Virus zu begegnen.

Sind Variationen vor diesem Hintergrund notwendigerweise ein Problem für die Wissenschaftler*innen? Variationen sind eine Herausforderung, aber sich mit ihnen auseinanderzusetzen, ist die Grundlage guter Wissenschaft. Denn sie zwingt Wissenschaftler*innen dazu, ihre Wahrheiten, Grundsätze und Überzeugungen zu hinterfragen, neue Probleme zu definieren und nach neuen Lösungen zu suchen, die uns Menschen helfen können, die Welt, die wir bewohnen, besser zu verstehen. Sie unterstützen uns dabei zu erkennen, wie wir auf eine friedliche, gesunde, dauerhafte und respektvolle Weise mit allen anderen Lebewesen zusammenleben können.

Dies ist ohne Variationen nicht möglich. Wie eben der vorliegende Beitrag, der die laufende wissenschaftliche Debatte innerhalb der Akademie widerspiegelt, zeigt: Variation in Form von interdisziplinärer Diskussion und Forschung unter Wissenschaftler*innen ist ein überaus hilfreiches Werkzeug für solche Überlegungen. Zwar ist, wie oben ausgeführt, der Austausch und die Zusammenarbeit zwischen den Disziplinen aufgrund der damit verbundenen Variationen (unterschiedliche Forschungsansätze, Theorien, Lehrmeinungen, Terminologien und Forschungsfelder) schwierig. Aber er ist ein höchst fruchtbares, produktives und notwendiges Unterfangen. Denn nur durch einen solchen Austausch, der die Gemeinsamkeiten, Unterschiede und Variationen zwischen den Materialien, wissenschaftlichen Methoden und Erkenntnissen der verschiedenen akademischen Disziplinen hervorhebt, sind die Wissenschaftler*innen in der Lage, produktiv über die Grundsätze ihrer Disziplinen zu reflektieren. Und das ist keine Kleinigkeit: Reflexion ist eine notwendige Voraussetzung für Kreativität, Qualität und Innovation in der wissenschaftlichen Forschung und Debatte und für ein besseres Leben auf der Erde.

3.   Transformation und Nutzen für uns Mitglieder

Wie in den obigen Ausführungen zu den Grundsätzen hervorgehoben wurde, besteht ein Bedarf an Kommunikation, Erklärung und Übersetzung in einem inter- und transdisziplinären Kontext. Dieses Bemühen beginnt auf der grundlegenden Ebene der Vorannahmen und der Ontologie (Grundsätze), umfasst unsere Analyseeinheiten (Variationen) und erfordert somit in der Folge eine Transformation von uns, den Wissenschaftler*innen. Wir müssen in der Lage sein, auf eine Art und Weise zu kommunizieren, die für Wissenschaftler*innen aus anderen Disziplinen verständlich ist, um „theoriefrei" zu sprechen, also abstrahiert von spezifischen theoretischen oder dogmatischen Paradigmen. Über eine enge disziplinäre Ebene hinauszugehen, indem wir unsere (disziplinären) Grundsätze und Variationen von einer Meta-Ebene aus reflektieren, kann uns erlauben, einen wirklich transdisziplinären Diskurs und eine transdisziplinäre Forschungspraxis zu etablieren. Dies setzt aus philosophischer Sicht voraus, dass eine Kommunikation zwischen Teilsystemen – in diesem Fall den hochspezialisierten Disziplinen in der Welt der Wissenschaft – möglich ist. Als Frankfurter Young Academy halten wir uns an dieses Habermas'sche Diktum.

Nur wenn wir in der Lage sind, miteinander zu sprechen und effektiv zu kommunizieren, können wir die Vorteile dieser Interdisziplinarität nutzen. Und diese Vorteile sind in der Tat lohnend: Durch den Austausch mit anderen Disziplinen können wir unser Problembewusstsein schärfen und von der Existenz dieser Probleme überhaupt erst erfahren. Ein Beispiel: Wenn ein oder eine Politikwissenschaftler*in der Goethe-Universität Frankfurt, die oder der sich für Fragen der Nachhaltigkeit interessiert, auf einen oder eine Wissenschaftler*in trifft, der oder die am Senckenberg Forschungsinstitut zur Tiefsee-Biogeographie forscht, können die Synergien überraschend und wegweisend sein. Die oder der Sozialwissenschaftler*in wird selbstverständlich nicht mit den hochspezialisierten wissenschaftlichen Daten und den Problemen bei der Datenerhebung und -analyse zur marinen Biodiversität vertraut sein. Dennoch wird ihr oder ihm das Verständnis der Schlüsselfragen, um die es geht, sehr helfen, nicht nur die politischen Probleme wahrzunehmen und zu verstehen, sondern auch ein Gefühl dafür zu bekommen, was evidenzbasierte Politikgestaltung in diesem speziellen Kontext bedeuten könnte. Ebenso könnte die oder der Wissenschaftler*in eine Ahnung davon bekommen, warum wissenschaftliche Erkenntnisse nicht direkt in politische

Lösungen umgesetzt werden, und begreifen, dass wissenschaftliche Erkenntnisse im politischen Machtspiel aller Wahrscheinlichkeit nach strategisch verwendet werden. Über den interdisziplinären Austausch hinaus könnte eine wirklich transdisziplinäre Perspektive sogar noch lohnender sein, wo das Überschreiten von disziplinären Grenzen uns erlaubt, neue Arten von Wissen zu generieren. Um beim Beispiel der Biodiversität zu bleiben: Die oder der Meeresforscher*in vom Senckenberg Institut ändert vielleicht die Wahrnehmung des Problems vollkommen, wenn sie oder er mit der oder dem Sozialwissenschaftler*in zusammenarbeitet, und umgekehrt. Eine integrative Sicht auf politische Probleme wie den Schutz der Biodiversität würde bedeuten, die unterschiedlichen Problemwahrnehmungen der einzelnen Disziplinen nicht nur zusammenzufassen, sondern sie zu überwinden und eine ganzheitliche Sichtweise einzunehmen.

4.   Transformation auf gesellschaftlicher Ebene

Letztlich ist also eine solche Transformation von uns selbst als Forscher*innen und Mitgliedern der Akademie förderlich für eine Transformation unserer Forschungspraxis; sie kann dadurch einen bedeutenden Beitrag zu umfassenderen Transformationsprozessen auf gesellschaftlicher Ebene leisten. Mit einer besseren und umfassenderen Erkenntnis dessen, welche Probleme am drängendsten sind, können wir etwas bewirken und werden auch in der Lage sein, den gesellschaftlichen Nutzen unserer Forschung einem breiteren Publikum effektiv zu vermitteln. Und hier schließt sich der Kreis: Was wir in der Young Academy gelernt haben, ist, allgemein verständlich über unseren hochspezialisierten Forschungsalltag zu sprechen. Der Austausch in der Young Academy hat uns geholfen, im öffentlichen Diskurs Position beziehen, mehrere Perspektiven einnehmen zu können und nicht an die engen Grenzen einer einzelnen Disziplin gebunden zu sein. Das ermöglicht uns, uns mit neuen und unerwarteten Perspektiven zu konfrontieren, so dass wir nicht jeder und jede für sich, sondern als Gruppe zum öffentlichen Diskurs beitragen und ihn sogar anregen können – dank der Akademie.

Interdisziplinarität, wissenschaftliche Freiheit und finanzielle Ressourcen ermöglichen es den Mitgliedern und Fellows der JQYA, sich vertieft mit wichtigen aktuellen Themen auseinanderzusetzen. Der transdisziplinäre Diskurs innerhalb der JQYA bietet nicht nur die Gelegenheit, relevante Themen umfassend aus verschiedenen Blickwinkeln zu beleuchten und einen Wissenstransfer unter uns Fellows und Mitgliedern anzuregen, sondern auch einen Transfer in die Gesellschaft. Daraus ergibt sich die Chance, aber auch die soziale Verantwortung, Impulse für die Gesellschaft zu formulieren. Wie kann die Akademie dieser Verantwortung gerecht werden, für sich und die Gesellschaft neue Impulse zu setzen?

Die Analyse unserer Grundsätze, Variationen und Transformation hat gezeigt, in welch hohem Maße unterschiedliche Expertisen in der Akademie zusammenkommen und wie daraus Transformation entstehen kann. Hier sehen wir die Chance, Themen von hoher gesellschaftlicher Relevanz, die der Beobachtung und Interpretation aus unterschiedlichen wissenschaftlichen Perspektiven bedürfen, für unseren transdisziplinären Diskursaustausch zu entwickeln bzw. aufzugreifen.

Wie könnte die Young Academy dies umsetzen? Welche Themen sollten aufgenommen werden? Und wie kann sich die JQYA Gehör verschaffen? Einige Beispiele aus den vergangenen Jahren können hier Vorbild sein. Sie zeigen, wie die Mitglieder der Young Academy wissenschaftliche und gesellschaftliche Debatten zu aktuellen Themen mitvoranbringen. Dies gelingt, wenn bei wissenschaftlich fundierten, publikumsfreundlichen Veranstaltungen gesellschaftlich relevante Themen aufgegriffen und beleuchtet werden. Was also bewegt die heutige Gesellschaft, wie kann die Young Academy zu diesen Debatten beitragen, wie hat sie es bisher getan? Drei Veranstaltungen der Young Academy haben globale Herausforderungen an Gesellschaft und Politik in öffentlichen Diskussionen aufgenommen und dabei reflektiert, welche Möglichkeiten und Probleme mit diesen Themen verbunden sind.

Covid-19-Pandemie und globale Gesundheit

Beginnen wir mit dem vielleicht aktuellsten Thema, der Corona-Pandemie. Wie stark Wissenschaft und globale Gesundheit gerade auch bei viralen Infektionen zusammenhängen, hat uns diese Pandemie deutlich vor Augen geführt. Biowissenschaftler*innen, Sozialwissenschaftler*innen, Ökonom*innen, Mediziner*innen und politische Entscheidungsträger*innen müssen aufs engste zusammenarbeiten, um in kurzer Zeit neue Projekte zu initiieren. Welche Chancen und Gefahren hier bestanden und bestehen, zeigte die öffentliche Diskussion von zwei Wissenschaftlern aus unterschiedlichen „Generationen" am 15.05.2020. Unser Member Christian Münch und Prof. Kai Simons erörterten verschiedene Aspekte davon, wie sich die weltweite Gesundheitssituation und der wissenschaftliche Fortschritt unter den Bedingungen der Virusinfektionen gegenseitig beeinflussen. Gleich zu Beginn der Pandemie präsentierte Christian Münch auf der Akademie-Plattform in einem öffentlichen Forum seine neuesten Arbeiten zur Definition der Antwort menschlicher Wirtszellen auf eine SARS-CoV-2-Infektion. Thematisiert wurden die gesundheitlichen Langzeitfolgen einer SARS-CoV-2-Infektion und die globalen Auswirkungen einer Pandemie auf alle Aspekte unseres Lebens. Der Gastsprecher des Forums, Prof. Kai Simons (Direktor emeritus des Max-Planck-Instituts für molekulare Zellbiologie und Genetik in Dresden), referierte über ein sich epidemieartig ausbreitendes und der WHO wohlbekanntes Gesundheitsproblem, die Adipositas. Er gab zu bedenken, dass die Adipositas unsere Gesellschaften in einer ähnlichen Art und Weise wie eine virale Epidemie untergrabe und langfristig negative

Auswirkungen auf die globale Gesundheit, Wirtschaft und sozialen Gefüge habe. Beide Sprecher eruierten Methoden einer personalisierten Gesundheitsberatung für ein besseres Leben.

Digitalisierung

Die Corona-Pandemie zeigte aber auch deutlich, wieviel Aufholbedarf in Deutschland bei der Digitalisierung von Schule, öffentlicher Verwaltung und anderem besteht und wie sehr dieser Ausbau aber mit Fragen von Grundrechten und Datenschutz einhergeht. Sicherlich stellt die Digitalisierung in der Wissenschaft nur einen Aspekt einer gegenwärtigen umfassenden Transformation dar. Wie die öffentliche Debatte von Nadine Flinner und Daniel Merk zeigte, die der Covid-19-Pandemie vorausging, hilft der Blick auf aktuelle Herausforderungen in der Forschung dennoch, die Möglichkeiten und Probleme dieses Wandels besser greifen zu können. An zwei angewandten Beispielen diskutierten sie am Academy Day (19.06.2020) den Einsatz der Künstlichen Intelligenz in medizinischen Anwendungen. Nadine Flinner beschrieb die Transformationen in den neuronalen Netzwerken und deren Variationen in verschiedenen Anwendungen der Krebsgrundlagenforschung. Daniel Merk befasste sich dagegen mit KI-unterstützter Medikamentenentwicklung. Sie beschrieben, wie dabei nicht nur technische Herausforderungen, sondern zwangsläufig auch ethische Fragen entstehen. Haben z. B. Patient*innen das Recht zu entscheiden, ob eine „Maschine" an ihrer Diagnostik beteiligt ist? Und wie geht man mit dem Risiko um, dass Ärzt*innen durch die Entscheidung der „Maschine" voreingenommen sind? Solche und ähnliche Fragen sind von größter gesellschaftlicher Relevanz.

Gender und Rolle der Frau

Gender, Gleichstellung und das Bewusstsein von struktureller Ungleichbehandlung von Geschlechtern sind ein weiteres zentrales Thema, das sich in den aktuellen gesellschaftlichen Debatten über gendergerechte Sprache, Frauenquoten, Me-too oder sexuell konnotierter hate speech niederschlägt. Auch in der Corona-Pandemie wurde oft gefragt, ob die Krankheit bzw. die Folgen ihrer Bekämpfung sich auf alle Geschlechter gleichermaßen auswirkt. Von einem Rückfall in alte Rollenbilder vor allem im Haushalt und bei der Betreuung der Kinder war und ist die Rede. Sicherlich trifft es zu, dass hier und überhaupt im Umgang mit beiden Geschlechtern historische Kontinuitäten zu beobachten sind. Gleichzeitig sind aber auch hier Variationen zu erkennen. Z. B. unterscheiden sich einzelne Länder stark in der Bewertung von männlichem und weiblichen Handeln. Dies zeigt sich besonders deutlich im Umgang mit Frauen in der politischen Öffentlichkeit. Dort begegnen Frauen besonderen, gender-spezifischen

Vorurteilen und Erwartungen. Welche dies sein können, und inwiefern sich hier Konstanten vom alten Rom bis in die Gegenwart ziehen lassen, diskutierten am 12.02.2021 die Fellows Sandra Eckert (Politikwissenschaft) und Muriel Moser (Geschichte) gemeinsam mit Prof. Sylka Scholz (Soziologie) vom Institut für Soziologie der Schiller-Universität Jena bei einem interdisziplinären Online-Panel, das die Young Academy gemeinsam mit dem Cornelia Goethe Centrum der GU, vertreten durch Prof. Helma Lutz, organisierte. Wie stark die Bewertung von weiblichem Handeln in der Öffentlichkeit von historischen Mustern geprägt ist, gleichzeitig aber auch lokalen Variationen unterliegt, wurde hier und in der lebhaften Diskussion der Teilnehmer*innen aus Wissenschaft und Gesellschaft deutlich.

Aber auch Fragen der Diversität sind in der Young Academy stets präsent. Hier vereinen sich Fellows und Members mit sieben verschiedenen Nationalitäten. Aktuell setzt sie sich aus acht Frauen und fünfzehn Männern zusammen, die bereits in den verschiedensten akademischen Traditionen gelernt und geforscht haben. Die Mehrheit ist derzeit an der GU tätig; eine Fellow ist aktuell Research Fellow am Aarhus Institute of Advanced Studies der Aarhus Universitet, ein Fellow arbeitet an der TU Darmstadt, zwei sind am Senckenberg Institut beschäftigt. Unter den Fellows/Members sind viele Familien mit (kleinen) Kindern. Familien werden finanziell unterstützt, die Veranstaltungen zu familienfreundlichen Uhrzeiten und mit Rücksicht auf die Schulferienzeiten geplant. Aber auch der intergenerationelle Aspekt ist in der Young Academy berücksichtigt. Die Zusammenarbeit von Fellows, Members, den Directors und den Distinguished Senior Scientists ermöglicht den Austausch zwischen verschiedenen Forscher*innengenerationen, der jedes Mal für beide Seiten aufs Neue äußerst fruchtbar ist. Schließlich regt die Young Academy durch ihre Veranstaltungen den akademischen Austausch auch außerhalb der eigenen Mauern an, mit Einrichtungen in Frankfurt – so der GU, dem Cornelia Goethe Centrum und dem Senckenberg Institut und der Wissenschaftlichen Gesellschaft –, aber auch mit anderen Forschungspartnern in der Rhein-Main-Region, so etwa der TU Darmstadt. Über ihre Distinguished Senior Scientists und weitere Veranstaltungen ist sie auch in der EU und in den USA präsent.

Betrachtet man diese Beispiele, so wird deutlich, dass die interdisziplinäre Debatte, die innerhalb der Young Academy möglich ist, einen wichtigen Motor für akademische Innovation darstellt. Sie wirkt auf der persönlichen Ebene von uns Einzelwissenschaftler*innen, aber auch auf einer übergeordneten Ebene innerhalb unserer akademischen Disziplinen und im weiteren Sinne für Frankfurt insgesamt. Damit ist die Young Academy ein wichtiger Beitrag zur Karriereentwicklung erfahrener Nachwuchswissenschaftler*innen in Hessen, der im bisherigen System der Forschungseinrichtungen gefehlt hat. Ihre Aktivitäten führen vor Augen, dass die Young Academy in Frankfurt, in der Rhein-Main-Region, in Europa und in den USA Forschung und gesellschaftliche Debatte anregt. Wie sehr dies der Fall ist, zeigt ihr Vermögen, etablierte Fürsprecher*innen für sich zu gewinnen. Nicht zuletzt wird dieser Beitrag durch ein

Vorwort von Herrn Prof. Herbert Zimmermann von der Wissenschaftlichen Gesellschaft Frankfurt eingeleitet, der vielleicht traditionsreichsten und renommiertesten wissenschaftlichen Gesellschaft Frankfurts. Möge diese wie auch unsere anderen Kollaborationen weiterhin fruchtbar und nützlich sein, und möge dies auch andere dazu anregen, sich unseren Debatten anzuschließen!

Award Ceremony 2020

ELENA WIEDERHOLD

June 19, 2020
Festsaal Casino, Campus Westend
Goethe University

At the Award Ceremony in 2020, we recognised eight early career researchers for their outstanding research achievements with admission to the JQ Young Academy. We were delighted to have such passionate and committed researchers joining us each year. New disciplines joined the JQYA this time: history, geography and bioinformatics.

The vice-president of Goethe University, Prof. Rolf van Dick, opened the ceremony and officially welcomed the new members as part of the JQ Young Academy family. In his welcome speech, Prof. van Dick emphasised the integrative character of the JQYA that brings together junior and senior scientists. Following up on this, Prof. Zimmermann, the president of the Scientific Society Frankfurt, demonstrated in his speech the importance of collaboration between generations. A great honour to the fellows and members and the culmination of the ceremony was the handing out of awards by Prof. Zimmermann.

Prior to the Award Ceremony, the fellows organised an Academy Day to launch the new annual topic 'Tenets, Variations, Transformations'.

JQ Young Academy welcomed in 2019–2020: Nadine Flinner (Bioinformatics), Philipp Erbentraut (Political Science), Muriel Moser (Ancient History), Torben Riehl (Marine Biology), Camelia-Eliza Telteu (Geography) as fellows; Benesh Joseph (Physics), Christian Münch (Biochemistry), Andreas Schlundt (Biology) as members.

Award Ceremony 2021

ELENA WIEDERHOLD

April 23, 2021
Renate von Metzler-Hall, Campus Westend
Goethe University

The third-year incoming fellows and members of the Johanna Quandt Young Academy were honoured on April 23, 2021, in a partly virtual Award Ceremony. By this event we had entered our third academic year and reached our maximum size. An important change has taken place in the past year: JQYA Founding Director Prof. Enrico Schleiff has taken up the post of president of Goethe University. We delightedly welcomed a new co-director Prof. Klement Tockner, the General Director of the Senckenberg Society.

The current president of the GU and former JQYA founding director, Professor Enrico Schleiff, welcomed the guests and pointed out that honouring young scientists by admitting them to the JQ Young Academy is not a simple membership application. Instead, it honours the candidates' outstanding achievements and commitment to science. After all, young scientists are the engine of innovative ideas.

Stefan Quandt, chairman of the Advisory Board of the Johanna Quandt University Foundation, welcomed the ceremony participants from the digital space and reflected on the pride of the fellows and members 'to be part of the international academic ecosystem' at Goethe University. In the name of his mother Johanna Quandt, her foundation and her legacy, Stefan Quandt wished all fellows a perfect time for both their academic career and their personal life.

Professor Matthias Lutz-Bachmann presented the new fellows and members their awards on behalf of Stefan Quandt and awarded three incoming Distinguished Senior Scientists, who will also contribute to this new cohort.

The programme's highlight was the keynote lecture, which was for the first time jointly written by all previous fellows and members. It was delivered by junior professor and political scientist Sandra Eckert.

JQ Young Academy welcomed in 2021: Sebastian Biba (Political Science), Tomás Cano (Sociology), Rikki Dean (Political Science), Susanne Fehlings (Ethnology), Philipp Dominik Keidl (Film and Media Studies), Magnus Ressel (History), Marco Tamborini (History and Philosophy of Science, TU Darmstadt) as fellows, and Cornelia Pokalyuk (Mathematics) as member.

As Distinguished Senior Scientists for the period 2021 to 2024, the new recruits are: Prof. Christiane Nüsslein-Volhard (Biology), Prof. Hartmut Rosa (Sociology) and Prof. Eleonore Stump (Philosophy).

Summer School 2020

ELENA WIEDERHOLD

August 3–4, 2020
Forschungskolleg Humanwissenschaften
Am Wingertsberg 4
61348 Bad Homburg

The programme of the first JQYA Summer School started with the topic introduction to 'Tenets, Variations, Transformations' where each participant shared spontaneous thoughts on this topic from the perspective of their own discipline. Followed by research seminars and round table discussions, a guided tour through Römerkastell Saalburg rounded off the two-day event.

Presentations and discussions:

Camelia-Eliza Telteu: 'How often do we think about climate change? What transformations do we need?'
Hanieh Saeedi: 'Marine biodiversity in a digital world'
Muriel Moser: 'Possibilities or Problems?'
Philipp Dominik Keidl: 'Gatekeepers of the Past: History Making as Industry and Fan Practice'
Nadine Flinner: TBA
Philipp Erbentraut: 'M. Ostrogorski (1854–1921) and the Transformation of the Democracy in America'
Sandra Eckert: 'Europe and COVID-19'

An Excursion into Roman Times
A Visit to the Saalburg Fort near Bad Homburg in the Taunus Mountains

FELIX KOTZUR

Guided tour by Felix Kotzur, a doctoral candidate in the Archaeology Department at the Goethe University of Frankfurt. Felix studies provincial Roman archaeology, and as an employee of the German Limes Commission, he has a profound knowledge of Roman antiquity in general and the monument inventory at Saalburg in particular.

As part of the Summer School on the theme of 'Tenets, Variations, Transformations' in August 2020, an excursion took the conference group to a special kind of sight in Hesse. Only a few kilometres from the venue in Bad Homburg, the Archaeological Park Roman Fort Saalburg awaited the participants on the crest of the Vordertaunus.

The Roman fort was part of the so-called Limes from the beginning of the 2nd century CE until the second half of the 3rd century CE. This was a complex border zone of the Roman Empire, which stretched over 550 kilometres from the Rhine to the Danube and separated the Empire from '*Germania*'. This borderline consisted of towers, wooden palisades or stone walls, ramparts and ditches, and over 100 smaller and larger military camps laid out in staggered rows.

The group met in front of the main gate of the camp, the so-called Porta Praetoria. Here, a few introductory words were said about the history of the site and its exploration. This began in the form of the first excavations in the 1850s under Gustav Habel and continued, sometimes more, sometimes less intensively, until the 1890s. The most important name in connection with the history of research is Louis Jacobi. He was a building councillor and architect in Bad Homburg. Thanks to his zeal and initiative, Emperor Wilhelm II – a frequent guest in the noble spa and casino town of Bad Homburg – became aware of the excavations. The emperor was enthusiastic about the monument from antiquity, which is a testimony to imperial power. He wanted to take up this tradition concerning his own German Empire by partially rebuilding this former military garrison and claiming it for himself. The reconstruction of the fortifications and some of the interior buildings were completed in 1907. Since then, the *Saalburg* has been a popular destination for excursions and the repository of several archaeological finds made along the Limes line in the Taunus.

The *Saalburg* was an excellent example of the theme of the Summer School, 'Tenets, Variations, Transformations', which was used to illustrate the aspects discussed in the lectures and discussions. The event established that scientific dogmas or principles are transformed by the factor of time, among other things. Structures, power and people are responsible for this.

The building stock of the *Saalburg* impressively shows how scientific principles change with the progress of research. Most of the buildings within the ring of walls are constructed so that the state of research in Wilhelmine times is considered appropriate. The façades are unplastered and show the bare quartzite masonry. In contrast, the newer buildings, such as the *Fabrica*, which opened in 2007, reflect the latest findings and are brightly plastered. In this case, structures that allow for continuous research condition these transformations.

An anecdote again highlighted another constellation of science and influencing factors. Here it became clear how power and individual persons could render science-based principles ad hoc null and void. In the reconstruction of the defence system, Louis Jacobi presented his designs to the emperor – as was customary. Wilhelm II almost always put his signature on it and let Jacobi have his way. However, when it came to the spacing of the battlements on the top of the wall, the emperor was dissatisfied and demanded narrower spacing, although Jacobi's wider spacing was based on original findings from Italy. A compromise was reached, which stated that the southern front, which the visitors saw first, would have the emperor's spacing between the battlements, and the rest would follow Jacobi's suggestion. The highest representative of the empire, and in his function also financier, prevailed. This, too, was noted earlier in the discussion among the participants: Science may enjoy – at best – independence in terms of methodology, but the aspect of funding is nevertheless essential.

IV. **Transformations in Debates
on Current Socially Relevant Topics**

Interplay of Science and Global Health in the Perspective of Viral Infections
Podium discussion

ELENA WIEDERHOLD / CHRISTIAN MÜNCH

May 15, 2020
Goethe University

Organiser: Christian Münch
Guest: Prof. emeritus Kai Simons, Max Planck Institute for Molecular Cell Biology and Genetics, Dresden

The most current topic, the Corona pandemic, was taken up in a forum on May 15, 2020. This pandemic has clearly shown us how strongly science and global health are interconnected, especially in the case of viral infections. Bioscientists, social scientists, economists, physicians and political decision-makers have to work closely together to initiate new projects in a short time. Which chances and dangers existed and exist here were revealed in the public discussion of two scientists from different 'generations'. Our member Christian Münch and Prof. Kai Simons discussed different aspects of how the global health situation and scientific progress influence each other under the conditions of the virus infections. Right at the beginning of the pandemic, Christian Münch presented his latest work on defining the response of human host cells to SARS-CoV-2 infection in a public forum on the Academy platform. Topics included the long-term health consequences of SARS-CoV-2 infection and the global impact of a pandemic on all aspects of our lives. The forum's guest speaker, Prof. Kai Simons (director emeritus of the Max Planck Institute for Molecular Cell Biology and Genetics in Dresden), spoke about a health problem that is spreading like an epidemic and is well known to the WHO: obesity. He suggested that obesity is undermining our societies in a manner similar to a viral epidemic, with long-term negative effects on global health, economies, and social fabric. Both speakers explored methods of personalized health counselling for better living.

Gender and Power: Women in the Political Public Sphere. Then and Now.
Scientific Tandem in Cooperation with
Cornelia Goethe Centrum für Frauenstudien und
die Erforschung der Geschlechterverhältnisse

SANDRA ECKERT / MURIEL MOSER

February 22, 2021
Organisers: Sandra Eckert, Muriel Moser
Guest: Prof. Sylka Scholz, Schiller University Jena

Gender, equality and the awareness of structural gender inequality are another central theme that is reflected in current social debates about gender-appropriate language, women's quotas, me-too or sexually connoted hate speech. In the Corona pandemic, too, questions were often asked as to whether the disease or the consequences of its control affects all genders equally. There was and is talk of a relapse into old role models, especially in the household and in caring for children. It is undoubtedly true that historical continuities can be observed here and in general in dealing with both sexes. At the same time, however, variations can also be seen here. For example, individual countries differ greatly in their assessment of male and female action. This is particularly evident in the way women are treated in the political public sphere. There, women encounter special, gender-specific prejudices and expectations. What these can be, and to what extent constants can be drawn here from ancient Rome to the present, were discussed on February 22, 2021, by Fellows Sandra Eckert (Political Science) and Muriel Moser (History), together with Prof. Sylka Scholz (Sociology) from the Institute of Sociology at Schiller University Jena at an interdisciplinary online panel organized by the Young Academy together with the Cornelia Goethe Centrum of the GU, represented by Prof. Helma Lutz. The extent to which historical patterns shape the evaluation of female action in the public sphere, but at the same time is also subject to local variations, became clear here and in the lively discussion among participants from academia and society.

Frauen in der politischen Öffentlichkeit – Damals und Heute

Veranstaltung in Kooperation mit dem Cornelia Goethe Centrum

SANDRA ECKERT / MURIEL MOSER

Online, einsehbar unter:
https://youtu.be/bZm_1hFzxRI?t=15612. Februar 2021.

Frauen in der politischen Öffentlichkeit begegnen besonderen, gender-spezifischen Vorurteilen, Erwartungen und Reaktionen. Welche dies sein können, und inwiefern sich hier Konstanten vom alten Rom bis in die Gegenwart ziehen lassen, war Thema einer Onlineveranstaltung mit Beiträgen aus der Geschichtswissenschaft, der Politikwissenschaft und der Soziologie. Die Veranstaltung wurde von der JQ Young Academy at Goethe in Kooperation mit dem Cornelia Goethe Centrum (CGC) ausgerichtet. Die geschäftsführende Direktorin des CGC Prof. Helma Lutz eröffnete die Veranstaltung mit einem Grußwort und einer Kurzvorstellung der drei Referentinnen.

Die Fellows der JQ Academy Dr. Muriel Moser (Geschichte) und Prof. Sandra Eckert (Politikwissenschaft) diskutierten mit Prof. Sylka Scholz (Soziologie) vom Institut für Soziologie der Schiller Universität Jena. Die Referentinnen erörterten in ihren Einzelbeiträgen sowie der sich anschließenden Diskussion, welche Muster bei der Bewertung von weiblichem Handeln in der Öffentlichkeit zu erkennen sind, wie diese historisch begründet sind, und wo Unterschiede nicht nur zwischen antiken und modernen Gesellschaften, sondern auch zwischen modernen Gesellschaften (hier das Vereinigte Königreich und Deutschland) auszumachen sind. Diese Fragestellungen wurden exemplarisch anhand von drei Beispielen untersucht, nämlich der römischen Aristokratin Clodia, der britischen Politikerin und ehemaligen Premierministerin Theresa May, und der zum Zeitpunkt der Diskussion noch amtierenden Bundeskanzlerin Angela Merkel.

Der öffentliche Angriff auf Clodia

Die Historikerin Muriel Moser eröffnete die Diskussion mit einem Vortrag über die römische Aristokratin Clodia Metelli. Ihr Vortrag beruht auf einer gemeinsamen Publikation mit Annika Klein (Klein und Moser 2020). Clodia, eine 34jährige Witwe, nahm im Jahr 56 vor Christus als Zeugin an einem öffentlichen Gerichtsprozess gegen Marcus Caelius Rufus, einem jungen Senator, teil. Bei diesem Prozess war Clodia die einzige Frau. Alle anderen Beteiligten – und es waren über 80 Personen aktiv daran beteiligt – waren Männer. Diesen Umstand machte sich Roms großer Redner Marcus Tullius Cicero, der den jungen Caelius verteidigte, zunutze. Er verwandelte den Prozess in einen Angriff auf Clodia und argumentierte, dass nicht Caelius, sondern Clodia schlecht gehandelt habe. Aus seinen Anschuldigungen lässt sich ablesen, welche Erwartungen an Frauen im öffentlichen Bereich gestellt, bzw. anhand welcher Rollenvorstellungen ihre Handlungen beurteilt wurden. Schnell wird klar, dass sich im alten Rom die Qualität einer weiblichen Handlung, und überhaupt einer Frau, stets daran maß, ob sie sich an männlichen Interessen orientierte. Eine gute Frau handelte dann gut, wenn sie dies in ihrer Rolle als Mutter, Tochter, Ehefrau oder Familienmitglied tat. Im Gegenzug konnten Frauen, die diese Rollen nicht erfüllten, bzw. deren Handlungen sich nicht in diese Beziehungen einreihen ließen, als moralisch verwerflich gelten.

Weil Clodia mit ihrem Lebenswandel keine diese Rollen ausfüllt, ja sogar ins Gegenteil verdreht, handele sie gemäß der Darstellung Ciceros wie eine Prostituierte – statt sich um ihre Aufgaben als Frau zu kümmern, gefährde sie mit diesem Prozess das Ansehen und die Karrieren guter Männer. Wir wissen nicht, ob oder wie Clodia auf die Worte Ciceros reagiert hat. In der Rezeption aber wirkt die negative Darstellung und Skandalisierung von Clodias Verhalten bis heute nach, wie durch zahlreiche Quellen belegt werden kann.

Die britische Premierministerin Theresa May –
Selbstdarstellung und Mediendiskurs

Die Politikwissenschaftlerin Sandra Eckert erörterte die übergreifenden Fragestellungen am Beispiel der ehemaligen britischen Premierministerin Theresa May. Grundlage für ihren Vortrag ist ein gemeinsam mit Prof. Charlotte Galpin (University of Birmingham) verfasster Buchbeitrag zu einem Herausgeberband über Frauen in Führungspositionen in der Europäischen Union und ihren Mitgliedstaaten (Eckert und Galpin i. e. 2021). Eckert und Galpin untersuchen sowohl die Selbstdarstellung der Amtsinhaberin sowie die mediale Darstellung und Bewertung ihrer Führungsrolle. Hierzu haben sie einen Textkorpus, bestehend aus 26 Reden Theresa Mays zum Brexit sowie aus mehr als 200 Zeitungsartikeln in britischen und deutschen Printmedien während der ca. dreijährigen Amtszeit der Premierministerin (13. Juli 2016 bis 24. Juli 2019), ausgewertet. Eckert bezog sich in ihrem Redebeitrag vor allem auf die Darstellung in den britischen Tabloids.

Zum Zeitpunkt ihres Amtsantrittes erörtern die Tabloids die Möglichkeit, dass Theresa May ihre Partei wieder zusammenführen, und außerdem die Verhandlungen zum Brexit erfolgreich abschließen könne. Vor allem aber wird ihr ausgefallenes Schuhwerk thematisiert, und ihr Führungsanspruch gegenüber der männlichen Parteiriege wird mit dem Ausspruch „Bei Fuß, Jungs" ironisch kommentiert. Hier führt Eckert zum Kontext von Theresa Mays Regierungszeit aus, dass es sich angesichts der mehrdimensionalen Krise – Brexit, Zusammenhalt des Vereinigten Königreiches, Spannungen in der Partei – um einen „glass cliff moment" handle. Mit diesem Begriff wird in der Literatur in Analogie zur Glasdecke („glass ceiling") eine besonders schwierige Ausgangslage bezeichnet, die eine weibliche Führungsperson fast schon zum Scheitern verurteilt, deren Scheitern aber dann auf Ihr Frausein zurückgeführt wird. Auch zum Ende der Amtszeit Theresa Mays wird auf eine stereotypische Rolle, nämlich ihre Rolle als Ehefrau, Bezug genommen: es soll ihr Ehemann sein, der ihr den klugen Rat gibt, endlich von der politischen Bühne abzutreten.

Die Autorität der deutschen Bundeskanzlerin Angela Merkel

Die Soziologin Sylka Scholz knüpfte in ihrem Vortrag an ihren umfangreichen Beitrag zur Geschlechterforschung, insbesondere zu Themen der Männlichkeitsforschung und Geschlechterbilder, an. Aus dieser Perspektive galt ihr Interesse unter anderem der Kanzlerschaft Angela Merkels (Scholz 2007, 2018), dem Schwerpunkt ihres Redebeitrages. Scholz konnte den Befund der Veranstaltung, dass Politik sowohl in antiken als auch in modernen Gesellschaften ein männlich konnotiertes Feld ist, in dem Frauen sich mit dem implizit männlich vergeschlechtlichten Leitbild eines Politikers auseinandersetzen müssen, weiter untermauern.

Scholz erläuterte in ihrem Vortrag das Konzept hegemonialer Männlichkeit und argumentierte, dass dieses grundlegend mit Macht und Herrschaft verknüpft, und insofern anschlussfähig an das Autoritätskonzept sei, wie es von den Philosophinnen Hilge Landweer und Catherine Newmark formuliert wurde. Vor diesem Hintergrund befasste sie sich in ihrem Vortrag mit der Frage, wie Angela Merkel während ihrer Amtszeit mit der sowohl institutionell als auch symbolisch verfestigten Verknüpfung von Autorität und Männlichkeit umging: Gelang es ihr, einen Anspruch auf Autorität zu formulieren? Wurde dieser Anspruch anerkannt? Scholz stellt auf der Grundlage ihrer medialen Analyse der textlichen Berichterstattung fest, dass es Merkel im Laufe ihrer Amtszeit gelang, ihren Anspruch auf Autorität weitgehend durchzusetzen. Auch visuell vermochte es die Kanzlerin, durch Rückgriff auf die etablierte politische Ikonographie Autorität und Weiblichkeit effektiv zu verknüpfen. Dies ließe sich, so Scholz, anhand der Porträts der Kanzlerin in ihrem Arbeitszimmer aufzeigen. Darauf aufbauend stellte die Referentin die Frage, ob die Kanzlerin eine neue Form „hegemonialer Weiblichkeit" etabliert habe. Dieser Begriff wird in der Geschlechterforschung genutzt, um die Wahrnehmung von Machtpositionen durch Frauen, sowie die damit verbundene Teilhabe an Gewalt- und Ausbeutungsverhältnissen, zu analysieren. Hier zieht sie das Fazit, dass derzeit von einem ambivalenten Zusammenhang von Weiblichkeit, Macht und Autorität zu sprechen sei; gleichwohl geriete das starre Verhältnis von Männlichkeit und Autorität in Bewegung.

Literaturhinweise

Eckert, S. und C. Galpin (i. E. 2021) Theresa May's leadership in Brexit negotiations: Self-Representation and Media Evaluations, in: *Pathways to Power: Female Leadership and Women Empowerment in the European Union*. Hg. I. Tömmel und H. Müller, Oxford: Oxford University Press.

Klein, A. und M. Moser (2020) Modern Scandal Theory and the Case of Clodia and Cicero in Ancient Rome, in: *Scandalogy*, Bd. 2, Cultures of Scandals, Scandals in Culture, Hg. v. A. Haller und H. Michael, Köln: von Halem Verlag, S. 82–103.

Scholz, S. Hg. (2007) *,Kann die das?' Angela Merkels Kampf um die Macht*, Bonn: Dietz Verlag.

Scholz, S. (2018) Die Autorität der Kanzlerin – Eine Annäherung, in: *Wie männlich ist Autorität? Feministische Kritik und Aneignung*, Hg. v. H. Landweer und C. Newmark, New York/Frankfurt: Campus, S. 31–55.

Establishing (a) Truths in Science
Scientific Tandem 'Tenets and Innovations'

PHILIPP ERBENTRAUT / MURIEL MOSER

At the Academy Day on June 19, 2020
Goethe University

Guest: Prof. Fleur Kemmers, Archaeology, Goethe University.

In many areas of science, irrevocable truths seem to prevail. Be it certain schools, topics or methods that form the specific canon of a discipline. These patterns are traditional tenets. In contrast, the panel examined the question of how new research approaches (innovations) can be implemented and propagated in science. The speakers provided insight into their own scientific practice – political science, history and archaeology – as well as the topics they are currently researching and how they came up with new ideas. The discussion with the audience then focused on the reactions young researchers experience when they challenge the dominant tenets and consider which framework conditions positively or negatively affect the acceptance of new approaches in research.

Digitalisation in Medical Science
Scientific Tandem 'Transformations in Neuronal Networks and their Variations in Different Applications'

NADINE FLINNER / DANIEL MERK

At the Academy Day on June 19, 2020
Goethe University

Science and society are constantly transforming, and these transformations influence each other. Digitalisation in science is only one aspect of a current comprehensive change. Using two applied examples, Daniel Merk and Nadine Flinner discussed artificial intelligence (AI) in medical applications. Nadine Flinner explained the basics behind neuronal networks, a modern tool which has gained increasing attention over the last few years in many areas of science. She elucidated what kind of transformations the measured data undergo to make valid predictions in various applications of basic cancer research. Daniel Merk, on the other hand, applies AI to drug development. The speakers introduced the kind of networks used in their research and explained what can be achieved using this modern technique. The speakers discussed with all the scientists present whether the application of this technique was interesting for their field or whether it was already used. Not only are there technical challenges, there are the inevitable ethical questions, too: do patients have the right to decide whether a 'machine' is involved in their diagnosis? And how does one deal with the risk of doctors becoming biased by the decision of the 'machine'? These fundamental questions remained open after a very lively and controversial discussion. One can imagine going into a public and political debate with these questions in future.

Climate Change in Relation to Water Management and Biodiversity
Scientific Tandem 'Tenets, Variations and Transformations Under and Above Water'

TORBEN RIEHL / CAMELIA-ELIZA TELTEU

At the Academy Day on June 19, 2020
Goethe University

This panel was jointly organized and dedicated to two different yet connected fields with 'water' as their common theme: abyssal hard rock under the water and the terrestrial water cycle above the water. Torben Riehl illustrated how his discovery of abyssal rock 'islands' transforms the current understanding of the evolution and distribution of abyssal biodiversity. According to this finding, a known homogeneous ecosystem changes towards a much more unknown complex ecosystem that supports a high heterogeneity of species composition, rocks, and habitats. On the other hand, Camelia-Eliza Telteu presented how global water models 'transform' from an array of complex sets of information into a simpler, unified suite of information simulating the terrestrial water cycle. Both speakers jointly summarised how climate change affects ocean biodiversity and the global water cycle, and opened the discussion on how climate change affects the ocean and the water cycle, as well as people's positive action towards climate change.

V. Other Scientific Contribution
Book Presentations

Our support line Scientific Development in the Framework of the Academy Programme contributes to all kinds of scientific formats, including book publications. In the last few years, three books and several different publications have appeared with the financial support of the JQYA. The book presentations were each written by the book authors themselves.

Florian Sprenger

Epistemologien des Umgebens. Zur Geschichte, Ökologie und Biopolitik Künstlicher Environments

(Edition Medienwissenschaft, 65), Bielefeld: transcript Verlag, 2019, 562 S. mit 37 s/w Abb. ISBN: 978-3-8376-4839-3

Das Buch wurde am 6. Juni 2019 im Rahmen einer Kooperation mit der Wissenschaftlichen Gesellschaft in Frankfurt Institut for Advanced Studies präsentiert. In seinem Vortrag präsentierte Florian Sprenger eine kurze Geschichte von Kreisen und Kreisläufen in den Diagrammen der Ökologie des 20. Jahrhunderts. Die Spannung zwischen den Harmonievorstellungen des Kreises und den Komplexitäten der Zirkulation machen deutlich, dass Ökologie – bis in die Gegenwart – immer Formen der Biopolitik enthält.

Der Aufstieg des Begriffs „Environment" zur Beschreibung der Gegenwart markiert den Einfluss, den das Nachdenken über Umgebungsrelationen und die Möglichkeit der technischen Gestaltung künstlicher Umgebungen seit Mitte des 19. Jahrhunderts gewonnen haben. In geschlossenen artifiziellen Welten wie Raumstationen oder künstlichen Ökosystemen wird die Verschränkung des „Environments" mit den umgebenen Organismen zum Gegenstand einer Biopolitik, die heute in autonomen Technologien der Umgebungskontrolle neue Räume erschließt. Florian Sprenger verfolgt diese Transformation ökologischen Umgebungswissens mit dem Ziel, gegenwärtige Technologien besser zu verstehen, den Begriff unselbstverständlich zu machen und die biopolitische Dimension jeder Ökologie herauszuarbeiten.

Federico L. G. Faroldi

Hyperintensionality and Normativity

Cham, Switzerland: Springer Verlag, 2019, 231 p., 117 b/w illustrations. ISBN 978-3-030-03486-3. ISBN 978-3-030-03487-0 (eBook)

The book was presented at a JQYA colloquium titled 'Ought, Good, Better: some Normative Themes, Formally' on May 5, 2019, at the Forschungskolleg Humanwissenschaften, Bad Homburg. In this colloquium, Federico Faroldi presented and discussed some themes from his book Hyperintensionality and Normativity (Springer, 2019), dealing with particular aspects of normative phenomena: deontic modality, practical reasons, and the relation between evaluative and descriptive/natural properties, leaving ample time for methodological remarks and open problems.

Presenting the first comprehensive, in-depth study of hyperintensionality, this book equips readers with the basic tools needed to appreciate some of the current and future debates in the philosophy of language, semantics, and metaphysics. After introducing and explaining the major approaches to hyperintensionality found in literature, the book tackles its systematic connections to normativity and offers some contributions to the current debates. The book offers undergraduate and graduate students an essential introduction to the topic, while also helping professionals in related fields get up to speed on open research-level problems.

Sandra Eckert
Corporate Power and Regulation. Consumers and the Environment in the European Union
(International Series on Public Policy), Palgrave Macmillan Cham, Springer International Publishing (Verlag), 2019, 354 p., 18 b/w illustrations, 1 illustrations in colour. ISBN: 978-3-030-05462-5. ISBN 978-3-030-05463-2 (ebook)

This book takes a fresh look at corporate power in the regulatory process. It examines how corporations seek to prevent, shape, make or revoke regulation. The central argument is that in doing so, corporations utilise distinct power resources as experts, innovators and operators. By re-emphasising the proactive role of business, the book complements our acquired knowledge of policymakers' capacity to put pressure on, or delegate power to, private actors. Empirically, the book covers European consumer and environmental policies, and conducts case studies on the chemical, paper, home appliance, ICT and electricity industries. A separate chapter is dedicated to the assumption that Brexit will lead to the unprecedented result of EU regulation being lifted, and how this could put corporate power in regulation at risk. This book provides a new perspective on the policy implications of corporate power to scholars, students and practitioners alike.

VI. Participants

The JQYA offers equal access to academics from a wide range of backgrounds, disciplines, nationalities, generations and genders. From the currently 24 fellows and members, every third person originates from a different country, nearly 40 % are female scientists, and every second scientist comes from a different discipline. With English as the working language, the JQYA connects young scientists with outstanding leaders from the international scientific community from the main fields of life sciences, humanities, social sciences, arts and literature.

Fellows

Fellows shape the inner core of the Academy. They are bound for three years to participate in all integral elements of the Academy Programme. They drive cross-disciplinary discussions within the Academy Programme and actively contribute to the organisation's goals and activities. Furthermore, fellows define the direction of the academic work by annually formulating the Academy Theme. In the first three rounds from 2018 until 2020, the JQYA gained 19 outstanding early career researchers from 11 disciplines affiliated with 14 departments.

Fellows shape the inner core of the Academy. They are bound for three years to participate in all integral elements of the Academy Programme. They drive cross-disciplinary discussions within the Academy Programme and actively contribute to the organisation's goals and activities. Furthermore, fellows define the direction of the academic work by annually formulating the Academy Theme. In the first three rounds from 2018 until 2020, the JQYA gained 19 outstanding early career researchers from 11 disciplines affiliated with 14 departments.

Sebastian Biba
JQYA Fellow 2020

Sabbatical Fellowship Award

Research area: political science
Research focus: China's foreign policy and international behaviour, China's international river politics, US-China-Europe triangular relations, responsibility in international relations

Vita
In a world without climate change and pandemics, Sebastian Biba would be an enthusiastic traveller. He loves hiking through nature, and he also loves exploring megacities. Hence, it used to be a good match for his personal interests that his research

focus has long centred on China and East Asia, where places of amazing nature exist next to buzzing metropolitan areas. While studying the Chinese language, history and politics in Beijing, Hong Kong and Taipei for more than three years, Sebastian toured many countries in East Asia and gained a deep understanding of their cultures and the extraordinary will of their people to get ahead. Today, he is certain: China and East Asia will shape the 21st century in profound ways; the only question that remains is, in what direction exactly?

In his academic research, Sebastian has mostly focused on analysing China's foreign policy behaviour in various contexts. He completed his PhD in Political Science at Goethe University Frankfurt in 2016 and has since been a senior researcher at the Chair for Political Science with a Focus on China/East Asia. Following his PhD, Sebastian held visiting positions at the School of International Studies (SIS) at Peking University in China and at the Institute for Asian Studies of the German Institute for Global and Area Studies (GIGA) in Hamburg. From September 2020 to June 2021, Sebastian moreover was a DAAD-sponsored visiting fellow with the Foreign Policy Institute of The Johns Hopkins School of Advanced International Studies in Washington, DC.

> The ancient Chinese philosopher Confucius is supposed to have said: 'There are three methods to gaining wisdom. The first is reflection, which is the highest. The second is imitation, which is the easiest. The third is experience, which is the bitterest.' As I strive after increasing my level of reflection, I am happy to be part of the JQYA family.

> The ancient Chinese philosopher Confucius is supposed to have said: 'There are three methods to gaining wisdom. The first is reflection, which is the highest. The second is imitation, which is the easiest. The third is experience, which is the bitterest.' As I strive after increasing my level of reflection, I am happy to be part of the JQYA family.

Current research projects

Sebastian is currently working on three larger research projects in correspondence with the various strands that make up his research interests. The first project is related to USA-China-Europe triangular ties. More specifically, Sebastian is interested in analysing current events around the emerging superpower rivalry between Washington and Beijing as well as what the resultant dynamics entail for Germany and the European Union. Sebastian's work is particularly focused on the question of how Germany and the EU should position themselves in light of growing Sino-American competition. Sebastian has already published a co-edited volume and a number of peer-reviewed journal articles and online commentaries related to this topic. He will later combine these publications to make up his habilitation.

The second project, for which Sebastian is currently drafting a research proposal to be submitted to the DFG for funding, deals with his long-term interest in China's shared river basins, especially the Mekong. Sebastian has established himself as one of the leading Western scholars in the field. In his upcoming project, he will examine the role of what he calls 'contending water-related infrastructures' for the balance of power as well as for inter- and intra-state conflict and cooperation in the Mekong River Basin.

Regarding his third ongoing research project, Sebastian seeks to break new ground in the field of international responsibility. Usually analysed in a normative context, Sebastian is more concerned with the empirical practice of responsibility fulfilment and responsibility denial. More precisely, Sebastian aims at finding out about the real-life conditions under which (state and non-state) actors tend to accept and deny responsibility attributions. For this purpose, Sebastian has been working on designing a new theoretical framework centred on the notion of 'responsibilisation'.

Eva Buddeberg
JQYA Fellow 2018

Sabbatical Fellowship Award

Research area: political science
Research focus: French philosophy and phenomenology, social and political philosophy, normative ethics/moral philosophy, methodology, critical theory, philosophy of language

Vita
Eva Buddeberg is assistant professor and a lecturer at Goethe University Frankfurt. She studied philosophy in Italy, France, the USA and Germany, and received her doctoral degree in 2009 from Goethe University Frankfurt. After that, she completed scientific stays abroad at the University of Chicago and London School of Economics. Eva was responsible for the German-French translation programme of the Éditions de la Maison des sciences de l'homme in Paris from 2006 to 2008. From 2008 to 2012 she was a research associate at the Cluster of Excellence 'The Formation of Normative Orders' at Goethe University Frankfurt. She is a member of Die Junge Akademie (Leopoldina) and of the Berlin-Brandenburgische Akademie der Wissenschaften. Her research interests are French and German philosophy and phenomenology, social and political philosophy, normative ethics/moral philosophy, methodology, critical theory and the philosophy of language. Eva is an extraordinarily experienced teacher, who designs BA and MA courses, drafts and marks essays and exams, chairs oral examinations and supervises BA and MA dissertations. She provides research advice and mentors pre-doctoral colleagues. Eva is frequently invited to give (keynote) lectures and talks.

Tomás Cano
JQYA Fellow 2020

Sabbatical Fellowship Award

Research area: social sciences
Research focus: family dynamics, gender inequality, child development, parenting strategies, social stratification

Vita
Tomás Cano is a research scientist at Goethe University Frankfurt. He is also affiliated with the Australian Research Centre of Excellence for Children and Families over the Life Course in Australia (ARC is similar to the European Research Council), and an Early Career Scholar 2020–2021 of the Work & Family Research Network in the United States. Prior to this, Tomás worked as a PhD candidate at the Pompeu Fabra University in Barcelona. He was a visiting scholar at the German Institute for Economic Research (DIW-Berlin), and the Universities of Queensland, Cologne and University College London.

> *Belonging to communities of researchers like the JQYA offer a wonderful opportunity to learn, improve, and share, together in spaces that are typically not common in research environments. It also allows for work interdisciplinarity, a key source of inspiration and advancement, due to the multifaceted background of the group of scholars participating at JQYA. These were fundamental reasons to be part of this excellent initiative.*

Current research projects
My most recent project analyses the role of the Covid-19 pandemic in gender inequality. In this project we are looking at key dimensions of gender equality that the pandemic might be altering so we can inform policy and practice for the implementation packages during the recovery years. The first of our focuses is on concerns: we looked at what concerns were creating distress in individuals, and how they vary by gender. The second focus is on single mothers and how their situations have changed during the pandemic, as this group is typically at the greatest risk of poverty and work-family conflict. And finally, we are also analysing how the pandemic has affected different dimensions of well-being for men and women, while also comparing changes in well-being across different recessionary periods (e. g., Covid-19, Great Recession).

Rikki Dean
JQYA Fellow 2020

Sabbatical Fellowship Award

Research area: social sciences
Research focus: democratic systems approaches, democratic innovations, digital democracy, participatory governance, political process preferences

Vita
Rikki Dean is postdoctoral fellow in the Democratic Innovations Research Unit at Goethe University Frankfurt. His current work is focused on: developing a systemic conception of democracy, evaluating participatory governance projects, understanding the impact of new online technologies on democracy, and analysing preferences for democratic governance. Rikki holds a PhD and an MSc from the London School of Economics, where his research focused on participatory innovations in the social policy process. His PhD thesis, Democratising Bureaucracy, received the LSE's Richard Titmuss Prize for Outstanding Scholarship. He also has a BA in Philosophy and Literature and an MA in Social and Political Thought from the University of Sussex. Rikki previously taught social policy at the London School of Economics and Democratic Innovations at the University of Westminster. He has worked on research projects for the LSE, Oxford University, the University of Manchester, the University of Birmingham and the University of Westminster. In 2015 he was a visiting democracy fellow at Harvard University's Ash Center for Democratic Innovation and Governance. Rikki has published numerous articles in international journals in political science and public policy. In 2018, he was awarded the Bleddyn Davies Prize from the journal Policy and Politics for his article Beyond Radicalism and Resignation: The Competing Logics for Public Participation in Policy Decisions.

> *Understanding a complex phenomenon like the democratic system requires bringing together insights and methods from an array of different fields of enquiry. This is what I try to practice in my work, drawing from political philosophy, public administration and political science, and it is why I am excited to join the Johanna Quandt Young Academy and engage with the diverse perspectives of the other fellows.*

Current research projects
Democracy: A Systems Approach
(with Brigitte Geißel and Jonathan Rinne)

Democracy is a contested concept that eludes a single specific definition and single institutional realisation. So, how should we theorise this elusive concept? With Brigitte Geißel and Jonathan Rinne, I explore how the systemic turn in democratic theory can enable us to theorise a conception of democracy that considers this heterogeneity. Drawing on a range of different traditions in democratic theory and quality of democracy scholarship and citizens' democratic preferences, we break down the root concepts that underpin different theories of democracy. The first publication from this project, 'Systematizing Democratic Systems Approaches', elaborated seven conceptual building blocks for thinking systemically about democracy: norms, functions, practices, actors, arenas, levels and interactions. We contend that this approach can help us to situate different democratic theories in relation to one another and enrich comparative political science on democracy.

Political Process Preferences in Europe: Rethinking Conceptual, Ontological and Methodological Foundations
Political scientists increasingly ask citizens about how they think society should be governed. However, in doing so, they almost universally adopt a model of democracy approach. They want to find out if you are a liberal democrat, a social democrat, a direct democrat, or even a stealth democrat. However, these models are neither psychologically nor sociologically realistic. Respondents are presented with abstracted and over-simplified choices that bear minimal resemblance to the complex, distributed decision-making of the political systems that they actually inhabit. This limits our ability to draw valid inferences from the responses to these questions. If we want to fully understand and address increasing dissatisfaction with democratic performance and growing authoritarianism in Europe, then we need a new approach to investigating process preferences. This project attempts to create and apply this new approach in order to rethink the conceptual, ontological and methodological foundations of the science of political process preferences.

Sandra Eckert
JQYA Fellow 2018

Sabbatical Fellowship Award

Research area: political science
Research focus: European integration, comparative politics, public policy

Vita
Sandra Eckert has been appointed full university professor of the chair in Comparative Politics at the Friedrich-Alexander University Erlangen-Nürnberg (FAU), starting her position in the academic year 2022–2023. She currently holds a three-year Marie Skłodowska-Curie COFUND Fellowship (October 2019 – October 2022) to conduct research as an associate professor at the Aarhus Institute of Advanced Studies (AIAS)

in Denmark. During her stay in Aarhus Sandra is on leave from Goethe University where she has been assistant professor (Juniorprofessorin) of politics in the European Multilevel System since October 2014. Sandra previously taught at the Universities of Berlin, Darmstadt, Freiburg, Mannheim and Osnabrück, and was a guest professor at Sciences Po Lyon. She worked as a research assistant within an EU-funded project at the Robert Schuman Centre of the European University Institute in Florence. Sandra received her PhD from the Free University Berlin, and graduated from the London School of Economics and Political Science, the University of Freiburg and the Université Paris 1 Panthéon-Sorbonne. She is the author of two books (Manchester University Press 2015, Palgrave Macmillan 2019), and has published in peer-reviewed journals such as the Journal of Common Market Studies, the Journal of European Public Policy, the Journal of European Integration, as well as Regulation & Governance.

> *I have experienced my time at the Academy as a unique opportunity and as a gift – an opportunity to exchange with excellent researchers from very different fields; a gift that gives us fellows extra freedom and independence in conducting our research. Holding a sabbatical fellowship allowed me to focus entirely on research and publication activities at a crucial stage in my career.*

Current research projects

Sandra Eckert's current research agenda centres around two key topics, the European Single Market and the Green Transformation in the European Union (EU). More specifically, Sandra studies the role of unelected bodies (e.g. supervisory and regulatory authorities) as well as the regulated industry in the process of integrating financial and energy markets. She was the principal investigator (PI) of the research project 'The State of the Union: The Politics of Integration in Banking and Energy' funded at the LOEWE Research Center SAFE (Sustainable Architecture for Finance in Europe), House of Finance, Goethe University Frankfurt, between January 2019 and December 2019. Since October 2019 Sandra has been conducting her own three-year research project "Everyday Life in the Single European Market. Consumer and Business Perceptions" at the Aarhus Institute of Advanced Studies, funded by the European Union and the Aarhus University Research Foundation. The project examines the perceptions of single market policies by consumers and business, but also their role in the regulatory process. Moreover, she is a principal investigator and member of the steering committee in the Jean Monnet Network VISTA "Revitalising the Study of EU Single Market Integration in a Turbulent Age", funded by the European Union for the period January 2019 to August 2023. The VISTA network seeks to promote new research and teaching on contemporary developments in the single market in the areas of defence, the digital market, finance and energy. Further, Sandra contributes to the research group "The Economic and Monetary Union at the Crossroads", financed by the LEIBNIZ Institute for Financial Research SAFE (Sustainable Architecture for Finance in Europe) during the period January 2019 to December 2022, with her research on reforms in the banking sector. On the topic of the Green Transformation, Sandra is a PI on the project "Regulating the Circular Economy in the OECD World" (REGCIRC) funded by the German-Israeli Foundation for Scientific Research and Development (GIF) as of January 2022. Moreover, she is currently researching and publishing on the EU's Green Deal, the transition towards a circular economy, and the role of business and finance in the EU's climate policy.

Philipp Erbentraut
JQYA Fellow 2019

Sabbatical Fellowship Award

Research area: political sociology
Research focus: political parties, elites and parliaments, comparative politics, history of democracy, political ideas in Germany, Great Britain and the USA in the 19th and 20th centuries

> *In my academic life, I have learned that nothing is more fruitful for one's own research project than engaging with the topics, perspectives and problems which other researchers face in their work. At the Johanna Quandt Young Academy, we can learn from each other. That is why I enjoy being a part of the great JQYA family.*

Vita

Philipp Erbentraut is a postdoctoral researcher at the Institute for Political Science at Goethe University Frankfurt. Philipp studied political science and history at the University of Greifswald and University of Bergen (Norway). During his research assistant position at the University of Düsseldorf and a lecturer position at the Seminar for Social Sciences at the University of Siegen, he graduated with a Dr. phil. from the University of Düsseldorf in 2015 with excellence. Philipp has been affiliated as a fellow with the postdoc-network 'Das Junge ZiF / The Young ZiF' for young researchers at the Center for Interdisciplinary Research (ZiF) with the University of Bielefeld since 2016. He was awarded with the Best Dissertation Award by the German Political Science Association (DVPW) in 2017 and a John F. Kennedy Memorial Fellowship by the Minda de Gunzburg Center for European Studies at Harvard University in 2018–2019.

Federico L. G. Faroldi
JQYA Fellow 2018

International Academy Fellowship Award

Research area: philosophy
Research focus: normativity and normative phenomena, hyperintensionality, philosophy of law, logic and metaphysics, formal ethics

Vita
Federico Faroldi is senior researcher of the Flanders Research Foundation (FWO) at the Centre for Logic and Philosophy of Science of Ghent University, Belgium. He received his international PhD degree in Logic and Philosophy from the University of Florence and the University of Pisa in 2017. Federico studied in Italy (Pavia), the UK (Oxford and St. Andrews), Ireland (Trinity College Dublin), France (ENS and EPHE) and the USA (NAU, NYU). In 2016–2017, he was an adjunct professor at the Department of Philosophy, University of Milan, where he taught topics in contemporary meta-ethics and metaphysics. Federico was part of the Nature and Normativity programme of the JQ Young Academy of Goethe University Frankfurt, where he was elected the International Academy Programme Fellow in 2018–19. Federico was also awarded a Lise Meitner two-year grant by the FWF, which he spent at the University of Salzburg (Austria) in 2019–20. In July 2019, he received his *abilitazione* as associate professor of logic and philosophy of science and as associate professor of philosophy of law in January 2020. Federico is the author of three monographs, The Normative Structure of Responsibility (London, 2014), Hyperintensionality and Normativity (Heidelberg/Berlin 2019), and Responsabilità e Ragione (Napoli, 2020), and about three dozen papers in international journals. He is co-editing a Springer Outstanding Contributions to Logic on Kit Fine with Frederik Van De Putte.

> *My research has been based on the idea of applying precise methods and formal techniques to normative matters and moves from conceptual analysis to formal methods to full-blown formal logic. Since my research is not confined to technical problems or language-related points, but confronts head-on substantial themes and problems, the multidisciplinary environment of the JQY Academy has been the perfect place to grow as a researcher.*

Current research projects
Current research projects are varied but are based on the idea of applying precise methods and formal techniques to substantive philosophical matters.

The first project, on generic reasoning, articulates the thesis that there is a kind of reasoning previously unrecognized as an independent type of reasoning. A theory of generic reasoning explains how a single significant instance may support general conclusions, with possible exceptions being tolerated. As a working hypothesis, this project will adopt that generic reasoning is irreducible to currently recognized kinds of 'pure' reasoning.

The second project, on Reason Structuralism, holds that (practical) reasons are identified by their place in a structure. The project aims to give a unified, fine-grained, precise account of the structure of reasons, which is able to explain their aggregation and double counting and to ask novel questions on their subtraction and partiality.

Susanne Fehlings
JQYA Fellow 2020

Academy Fellowship Award

Research area: ethnology
Research focus: economic activities of small and medium entrepreneurs, shuttle traders in post-Soviet Eurasia, interethnic exchange between Armenian, Russian, Georgian and Chinese traders and businesspeople

Vita
Before Susanne Fehlings came to Frankfurt, she studied archaeology, history of art and social and cultural anthropology in Paris, Moscow and Tübingen. Since 2016 she has been the leader of a Volkswagen Foundation-funded research project on 'informal markets and trade in Central Asia and the Caucasus'. As an anthropologist she has spent a lot of time conducting fieldwork abroad – in Russia, Ukraine (Crimea), Central Asia and the Caucasus. For Susanne, science means the encounter with the unknown. She is driven by the interest in understanding alien societies and cultures holistically, as embedded into broader and global contexts. In her leisure time, Susanne is a passionate reader, hobby artist and a member of Naturschutzbund, the German Nature and Biodiversity Conservation Union.

Nadine Flinner
JQYA Fellow 2019

Academy Fellowship Award

Research area: computational and medical sciences
Research focus: correlation between cell morphology, function and the underlying molecular features, link between features and morphology, deep learning and quantitative image segmentation, artificial intelligence in medical science

Vita
Nadine Flinner is a research fellow at Frankfurt Institute for Advanced Studies (FIAS). Nadine studied bioinformatics and worked on the structure and phylogeny of membrane proteins during her diploma thesis. In her doctorate, which she completed in 2015, she investigated the behaviour of membrane proteins using molecular dynamic simulations. Nadine started to study the migration of immune cells as a postdoc at FIAS and is now working on understanding the correlation between cell morphology and the underlying molecular characteristics.

Philipp Keidl
JQYA Fellow 2020

Academy Fellowship Award

Research area: film and media studies
Research focus: fan cultures, production of nonfiction media, fan-made documentaries, film and media history, justice system, environmental crises

Vita
Philipp Dominik Keidl is a postdoctoral researcher in the Deutsche Forschungsgemeinschaft Graduate Training Program 'Configurations of Film' at Goethe University Frankfurt. His research on media fandom and moving image heritage is published and forthcoming in The Journal of Cinema and Media Studies and The Journal of Popular Culture and Film Criticism. He is currently finishing his monograph Plastic Heritage: Fans and the Making of History, which examines fandom as a participatory historical culture. Philipp holds a PhD in Film and Moving Image Studies from Concordia University in Montreal and an MA in Preservation and Presentation of the Moving Image from the University of Amsterdam.

Daniel Merk
JQYA Fellow 2018

Academy Fellowship Award

Research area: pharmaceutical and medicinal chemistry
Research focus: medicinal chemistry and pharmacology of nuclear receptor ligands, chemical probes and tools, artificial intelligence for molecular design

Vita

Daniel Merk has been a W3 professor and chair for pharmaceutical and medicinal chemistry at Ludwig Maximilian University of Munich since 2021. After graduating in Pharmaceutical Sciences and Pharmacy at LMU Munich, Daniel moved to Goethe University Frankfurt in 2011, where he received his PhD degree in Pharmaceutical and Medicinal Chemistry in 2015. Between 2015 and 2017, he was an early career group leader at Goethe University. In 2017, he obtained an ETH fellowship and conducted research at the Swiss Federal Institute of Technology (ETH) Zurich for two years before returning to Goethe University for his habilitation in 2019. Between 2019 and 2021, Daniel was a group leader at the Institute of Pharmaceutical Chemistry at Goethe University Frankfurt. He won a Heisenberg grant from the German Research Foundation (DFG) and was offered a full professor position by the University of Gothenburg in 2020 but accepted an offer by LMU Munich to become a pharmaceutical and medicinal chemistry chair in 2021. Daniel has been awarded multiple prestigious grants and prizes, including an ETH Fellows Scholarship, the Phoenix Pharmaceutical Sciences Award in Pharmaceutical and Medicinal Chemistry (2018), the Aventis Life Science Bridge Award (2019) and the Innovation Award in Pharmaceutical and Medicinal Chemistry by the GDCh and DPhG (2020).

> *I have always loved the Pharmaceutical Sciences for their interdisciplinarity in uniting many natural sciences. The JQYA is enjoyably multidisciplinary and inspiring, too.*

Current research projects

Daniel's interdisciplinary research focuses on the design of chemical tool compounds to control biological processes. It currently addresses the ligand-sensing transcription factors retinoid X receptor (RXR), nuclear receptor-related 1 (Nurr1) and tailless homologue (TLX). These proteins seem to be highly involved in neurodegenerative diseases such as Alzheimer's disease, Parkinson's disease and multiple sclerosis. Pharmacological modulation of RXR, Nurr1 or TLX might hold neuroprotective and even neuro-regenerative potential, thereby opening new avenues in neurodegenerative disease treatment. However, drug-like molecules selectively binding to and modulating these transcription factors with high affinity are lacking but needed to evaluate pharmacological control of RXR, Nurr1 and TLX in neurodegeneration and beyond. Daniel's group is developing selective high-affinity tool compounds to activate or block the transcription factors. These chemical tools enable pharmacological studies to evaluate and validate the therapeutic potential of RXR, Nurr1 and TLX. By using these custom tools in functional studies, Daniel also aims to obtain a molecular and mechanistic understanding of RXR, Nurr1 and TLX modulation by small molecules, including cellular interaction partners, signalling pathways and phenotypic consequences. In addition, Daniel studies generative deep learning models for molecular design. Such machine learning algorithms can be trained on 'raw' molecular representations and enable rule-free, data-driven de novo design of bioactive molecules. Daniel studies how such models can be used to access desired regions of the chemical space and inspire and accelerate early drug discovery.

Muriel Moser
JQYA Fellow 2019

Sabbatical Fellowship Award

Research area: ancient history
Research focus: political and cultural history of the Graeco-Roman world, especially late-antiquity emperors and senates, Athens in the Roman empire, and women in Roman culture

Vita

Muriel Moser is assistant professor of ancient history at Goethe University Frankfurt. She has been working and teaching in Frankfurt since 2012. Muriel studied Classics at the University of Cambridge (BA, MPhil), where she graduated with distinction. After an academic sojourn at the ENS-Rue d'Ulm in Paris and a scholarship at the Institute of Ancient History of the University of Cologne, she graduated with a PhD in Classics (with distinction) from the University of Cambridge in 2013. She has been working and teaching in Frankfurt since 2012. From 2015 to 2018, she was a sub-project leader in the DFG-funded SFB 1095 'Schwächediskurse und Ressourcenregime'. Muriel was awarded multiple scholarships at the University of Cambridge, as well as a scholarship of the Alfried Krupp von Bohlen und Halbach-Stiftung at the University of Cologne, a Karl-Ferdinand-Werner-Fellowship from the German Historical Institute (DHI) in Paris, and a sabbatical grant from the Forschungszentrum Historische Geisteswissenschaften (FZHG) in Frankfurt. Her first monograph, 'Emperor and Senators in the Reign of Constantius II', was published by Cambridge University Press in 2018. Her publications (in English, German and French) also include a thematic volume and several papers in international peer-reviewed journals, a collective volume on Roman Athens, and several other papers and reviews. Muriel has organized a number of international conferences and workshops. She works as a peer reviewer for several publishing houses, journals and institutions.

> My ongoing fascination drives my academic career for the political and cultural developments of Greek and Roman antiquity. I joined the Johanna Quandt Young Academy because it offers a platform for interdisciplinary academic discussion and outreach activities in Frankfurt and beyond.

Current research projects

My current research project investigates political uses of the past in Athens in late Hellenistic and early Roman times (100 BC to AD 100). In this monograph, provisionally entitled, 'The Politics of the Past in Athens from Ariston to Plutarch', I examine texts, inscriptions and archaeological sources to reconstruct how Athenians and Romans used references to Athens' past to further their political aims in a time of great political upheaval. In so doing, I also disentangle the web of modern assumptions about the relationship between Athenian culture and Roman power. Thereby, my study advances our knowledge of the history of Athens and her relationship with Rome in a crucial period of her history. At the same time, it provides new insights into the role of Athens and Athenian history in Roman politics from Sulla to Nero, that is, during the civil wars of the later Roman Republic and the end of the first imperial dynasty of Rome, the Julio-Claudians. In this way, it can contribute to several current research debates about the nature of Greek and Roman culture and politics in this period.

I am also interested in the role of women in Roman politics and society. In this field, I have worked in particular on the scandalization of female presence in the public domain of the Roman Republic. Together with Dr Annika Klein, I examined the scandal of Clodia in Cicero's forensic speech Pro Caelio. Our research argues that modern scandal theories can help us better understand the dynamics of scandalization in ancient Rome and present societies. In this context, I organized the online panel 'Frauen in der politischen Öffentlichkeit' together with Sandra Eckert, where we discussed the treatment of women in the public domain in ancient Rome and modern European democracies.

Magnus Ressel
JQYA Fellow 2020

Academy Fellowship Award

Research area: early modern history
Research focus: cultural history, social and economic history, infrastructural history, Industrious/Industrial Revolution, Technological Enlightenment, transatlantic slave trade

Vita

Magnus Ressel studied history, philosophy, anthropology and business administration in Saarbrücken and Sydney. After his undergraduate studies, he became a PhD student in Munich. He completed his thesis on the relationship between Northern Europe and Northern Africa in the early modern age as a French-German 'cotutelle' at the universities of Bochum and Paris 1 Sorbonne, with 'summa cum laude' and 'très honorable avec les félicitations du jury'. Following that, Magnus spent one year as a fellow of the Alexander von Humboldt-Foundation in Padua to study the German merchant communities in Venice and Livorno during the 18ᵗʰ century. In 2013, he briefly became an assistant professor at the Rijksuniversiteit Groningen, before opting for the University of Frankfurt. Here, he was coordinator of the International Graduate School 'Political Communication from Antiquity to the 20ᵗʰ Century' and assistant professor at the chair of Early Modern History (2013–2019). He left Frankfurt for over a year to finish his habilitation at the German Historical Institute Rome and the Institute for Advanced Study in Munich in 2015–2016. Since 2020, Magnus has been a fellow of the Gerda Henkel Foundation. He intends to bring a historical perspective to the JQYA that focuses on intercultural contacts and transfers. Methodologically, he mostly takes inspiration from a global history, institutional theory, historical anthropology and network research – enriched by the tools of digital humanities.

> *The most exciting adventure that we can undertake in our age is scientific research – in my case, in the archives. As a historian of the early modern age (1500–1800), I look at human societies at the threshold of modernity. My greatest motivation to be actively involved in the JQYA is to be part of an interdisciplinary team that strives together for new knowledge and concepts.*

Current research projects

Based on an extensive company archive, dispersed over Bordeaux, Paris and Brussels, I currently study the global enterprises of Friedrich (since 1784 von) Romberg (1729–1819). He was probably the wealthiest entrepreneur on the European continent around 1785 because of his intensive involvement in the Belgian textile industries, banking, maritime insurance, the Atlantic colonial and slave trade, and freight forwarding from Ostend down to Southern Italy and the Balkans. Operating in Brussels under the direct protection of the Emperor, Romberg vertically monopolised various firms' diverse sections and aspects of the exchange and refinement of continental European products and colonial goods. He was thus able to maximise his profit margins in the most profitable business fields of the late 18ᵗʰ century. Romberg's activities stand at the threshold of the historical change around 1800. Among the more important businessmen, we see the figure of the 'classical' merchant banker increasingly disappearing during these years. Instead, these figures turned into the directors of institutionalised companies with specialised subsidiaries. In Romberg's biography, we see this change taking place in the trajectory of one person.

By looking at Romberg and his various companies, the project aims to add new layers to a fundamental debate of economic history that is of great importance, especially in the history of European Atlantic colonialism. This is the transformation of traditional trading companies, organised around core families, into formalised and bureaucratised institutions in the decades around 1800. In a classic thesis from 1944, Eric Williams has put forward the idea of the profits of Atlantic plantation slavery as a necessary precondition for the Industrial Revolution through the reinvestment of capital accumulated in colonial businesses. The debate on this has not subsided yet. I intend to give more nuance to this debate and thus enhance the sophistication of the underlying model. I believe that the more intense economic entanglement of the European hinterland and the colonial world increased the complexity of business to such a degree that this necessitated the transformation of the herein engaged companies towards a more institutionalised structure. In addition to substantial capital accumulation, new challenges in organising complex business structures caused the fundamental and soon irreversible transformation of companies to more modern structures in the decades around 1800. Research into the business structures of Romberg's business empire thus promises a new understanding of a threshold of economic and corporate history.

Torben Riehl
JQYA Fellow 2019

Academy Fellowship Award

Research area: marine geobiodiversity
Research focus: marine species discovery and extinction-risk assessment, origins of deep-sea biodiversity, drivers of diversification at depths > 3500 m, seafloor habitat heterogeneity

Vita
Torben Riehl is a research staff member, junior research group leader, science communicator and deputy head of the section Crustacea at the Senckenberg Research Institute and Natural History Museum Frankfurt. His first encounter with marine science was in 1998 during an internship at the Institute for Coastal Research at the Helmholtz-Zentrum Hereon (formerly GKSS Research Centre). Later, Torben studied biology, majoring in zoology as well as hydrobiology and fisheries science at the University of Hamburg, where he also conducted his doctoral studies and graduated in Zoology with the highest distinction as Dr. rer. nat in 2014, supervised by Prof. Dr Angelika Brandt. During his doctoral studies, Torben was awarded a PhD fellowship of the German Academic Scholarship Foundation and the Geddes Visiting Research Fellowship of the Australian Museum, Sydney, where he spent about 2.5 years between 2011 and 2013. During 2014–2017 Torben worked in two BMBF-funded research projects dedicated to deep-sea biodiversity in the central Atlantic and Northwest Pacific. After this postdoctoral period at the Centre for Natural History Hamburg and in the Marine Biology Department of Ghent University, Torben received a tenure position at the Senckenberg Research Institute Frankfurt and started his habilitation at Goethe University in 2017. Since his biology studies in Hamburg, Torben has repeatedly been on campaigns spanning weeks with research vessels studying deep-sea biodiversity.

His scientific contributions include, amongst others, species discoveries and new taxonomic descriptions, speciation studies and deep-sea habitat mapping. Besides his JQYA fellowship, he is also a member of the Junge Akademie | Mainz of the Academy of Sciences and Literature Mainz.

> *Becoming a researcher was a childhood dream of mine. Curiosity for nature, especially underwater life, has fascinated me since I was very young. Growing up in the glacial valley of the Elbe River sparked my curiosity in the exploration of 'unknown' (at least to me at that time) aquatic biodiversity. Since then, it has been my primary interest and has provided the cornerstone for my career in marine biology.*

Current research projects

In 2021 Torben raised philanthropic 'kick-off' funding and founded the 'Senckenberg Ocean Species Alliance' (SOSA) together with Prof. Dr Julia Sigwart. With this project, Torben is establishing and coordinating a research service unit dedicated to accelerating the discovery of ocean species, their conservation, and enhancing public appreciation and valuing of the oceans' species. SOSA follows a combined research-conservation-communication strategy, for which aesthetic (macro-) photographic images of marine species played a key role. Imaging techniques, such as photography and confocal laser-scanning microscopy, are an important aspect of Torben's research and communication, generously supported by the JQYA fellowship.

The DFG-funded project 'Rocky seafloor – Underestimated driver for Benthic Biodiversity Living in the abyss and its Evolution' (RUBBLE) was accepted in August 2021. With this project, Torben will investigate links between habitat heterogeneity and biodiversity near bedrock patches in the abyssal Atlantic – as the PI of the project and chief scientist of a > 26-day sampling campaign with RV METEOR. The RUBBLE project will record the fauna of sediments adjacent to isolated bedrock patches investigating the dependence of biodiversity patterns and genetic differentiation on habitat heterogeneity. By quantifying abyssal biotopes and the habitat-type dependent modelling of biodiversity over bathymetrically mapped areas, RUBBLE lays the foundation for hydroacoustic-based biodiversity assessment and modelling for environmental protection and environmental planning abyssal areas.

The project 'Effects of Depth, Distance, and the Geomorphology on deep-sea Isopod differentiation' (EDDGI) was initiated in 2021 following a pilot study supervised by Torben. In this project, Torben aims to use integrated taxonomic, and population genomic research approaches on extensive samples from the Northwest Pacific deep sea to revisit the question of the ecological and evolutionary origins of deep-sea biodiversity, and also to answer innovative and forward-looking research questions and communicate them to the general public. In this context, the JQYA funded a three-month research assistant to prepare a manuscript based on the pilot study. A DFG proposal based on the pilot study has been submitted for review.

Hanieh Saeedi
JQYA Fellow 2018

Academy Fellowship Award

Research area: marine science
Research focus: ecological and evolutionary process of biodiversity patterns, biogeography of marine species (shallow and deep-sea), species richness and distribution
shift under future climate change, big datasets management, Intergovernmental Science-Policy Platform on Biodiversity and Ecosystem Services

Vita

Hanieh Saeedi has been the biodiversity information coordinator at the Senckenberg
Research Institute and Natural History Museum in Frankfurt since 2020. From 2017 to
2020 she was a postdoctoral researcher at Goethe University Frankfurt and Senckenberg Research Institute and Natural History Museum, a fellow of the JQYA at Goethe
University Frankfurt, and a data manager in the OBIS Deep-Sea Node of UNESCO.
Hanieh studied for an M.Sc. in Marine Biology at the Shahid Beheshti University, Iran
(first ranked student), and graduated with a PhD in 2015 from the University of Auck-
land, New Zealand. Her research focuses on the digitisation of museum collections,
biodiversity informatics (using big data), and uncovering the species distribution
patterns under future climate change. For this, Hanieh uses ecological modelling and
provides fundamental information and assessments on the global status of biodiversity in the world's oceans in response to the request of policymakers to better establish
management plans to maintain ocean biodiversity.

> Since I was a child, I have always been fascinated with the beauty and calming effect of the ocean
> in my life. As I grew up, I learned more about the underwater wonderland and dreamt about
> being a marine scientist one day. My understanding and perception of science and research have
> changed over time as my career developed internationally and interdisciplinary. The Johanna
> Quandt Young Academy is very interdisciplinary; we learn from each other. That is why I enjoy
> being a part of the great JQYA family.

Current research projects

I am interested in understanding the driving factors (ecological and evolutionary processes) that shape biodiversity patterns and biogeography in marine species (shallow
and deep-sea) using big data. In addition, I am interested in predicting how these biodiversity patterns and species distribution ranges will change under future climate
change. I am also the data manager for OBIS (Ocean Biogeographic Information System) deep-sea nodes at UNESCO, specialising in large dataset management, biodiversity data standards and quality control tasks. To conduct my research, I use different
skills and apply different methods and techniques, such as taxonomy (morphology

and molecular biology), biogeography, biodiversity informatics, macroecology, and ecological niche modelling.

I currently lead projects in museum collection digitisation, biogeography, biodiversity informatics using big data at regional (e. g. NW Pacific) and global scales. My research is mainly based on natural history collections and the digitised open data. My long-term visions rely on digitised products rather than the physical specimen. Creating digital-only workflows facilitate digitisation, curation, and data linkages. Thus, this preserves the value of physical specimens, establishes a global community, and develops automated approaches to advance biodiversity discovery and conservation. These efforts will transform large-scale biodiversity assessments as a response to fundamental questions brought by policymakers, including those concerning the contemporary issues of global change.

I also work for science-policy intergovernmental bodies such as IPBES (Intergovernmental Science-Policy Platform on Biodiversity and Ecosystem Services) to provide baseline information for biodiversity assessment and marine invasive species reports in response to policymakers to better understand the global status of changing biodiversity and Anthropocene biodiversity loss in the world's oceans and consequently establish more effective strategic management plans for ocean biodiversity conservation and sustainability.

In 2020, I was nominated and elected as a member of the German UN Ocean Decade Committee, which aims to make the UN Decade of Marine Research for Sustainable Development known in Germany, to support its implementation in Germany and act as a link between national and international activities. My vision in this committee is to implement transformative solutions across different disciplines towards the protection and sustainable use of the ocean.

JQYA has been very supportive during the last few years by providing financial and academic support to enable me to carry out and continue my research and provide a pleasant environment with other members in a very international and interdisciplinary atmosphere, where we can discuss and learn from each other.

Peter W. Smith
JQYA Fellow 2018, JQYA Alumnus since 2021

Academy Fellowship Award

Research area: linguistics
Research focus: syntax and morphology, particularly the syntax and morphology of agreement, number, and case

Vita

Peter W. Smith has been a senior technical writer at Wärtsilä Netherlands since March 2021. Peter studied linguistics at University College London, where he received his BA in Linguistics with Honours. He then studied at the University of Connecticut and there successfully defended his PhD thesis Feature Mismatches: Consequences for Syntax, Morphology and Semantics in 2015. Between 2015 and 2020, he was a postdoctoral researcher at Goethe University Frankfurt. There, he continued his collaborative work on suppletion and headedness in Distributed Morphology with Beata Moskal. Peter also published work on grammatical roles and their role in modern linguistic theory at this time. In 2021 his book 'Morphology – Semantics Mismatches and the Nature of Grammatical Features' was published by De Gruyter. In 2021 he left linguistics and the academic world behind after becoming disillusioned with the current state of his field, the associated career prospects, and academic life in general, and decided to pursue a career with a more normal work-life balance.

> Being a member of JQYA was a great opportunity for me. I loved being part of an interdisciplinary group and learning from others about their research. Researching only linguistics was very narrow, and so it was stimulating to hear work from other fields. It was especially nice to be around others at a similar stage in their careers.

Research projects

Peter worked on various topics as a researcher in the areas of morphology and syntax. He worked on three broad topic areas. The first was so-called 'hybrid nouns', nouns whose surface shape do not match their interpretation (e. g. Mädchen, which is grammatically neuter but semantically feminine), and the patterns of agreement that they show with verbs. It is possible to distinguish between an agreement that is motivated by the semantics of the noun versus one that is determined by the morphological shape of the noun. These two types of agreement show different properties and proved to be insightful for how agreement is modelled in syntactic theory.

The second topic Peter worked on, most prominently with Beata Moskal, but also Jungmin Kang, Ting Xu and Jonathan Bobaljik, was patterns of suppletion in the world's languages, suppletion being the phenomenon where a drastic irregular change is seen in the shape of a word when inflecting for a grammatical value, for instance, long–longer–longest vs good–better–best. Their findings showed, in congruence with other work from Beata Moskal and Jonathan Bobaljik, that suppletion is not a free phenomenon but rather obeys clear rules that determine where it is possible and where it is not possible.

Finally, Peter worked on the role that grammatical rules play in modern syntactic theory. Intuitively a fundamental part of how we use language, the terms subject, object, etc. play a minor role in prominent theories of generative syntax that follow the work of Noam Chomsky, a situation that has been true since the 1970s. Peter reinvestigated this topic to try to determine whether this viewpoint was still sustainable in light of the evidence gathered in the intervening decades that suggest there should be a role. He left the field before the investigation was complete, but work published as part of this project seemed to suggest that whilst it was possible to handle these newer cases without assuming that grammatical functions play a role, it is not nearly as easy, or insightful, as Chomskian work would like it to be.

Florian Sprenger
JQYA Fellow 2018, JQYA Alumnus since 2019

Sabbatical Fellowship Award

Research area: media studies
Research focus: media histories and media concepts, history and present of digital cultures, infrastructures of the Internet, biopolitics of artificial environments, media ecologies and epistemologies of the environment, transport transformation and autonomous cars, smart urbanity and the internet of things

Vita
Florian Sprenger has been a professor of virtual humanities at the Institute for Media Studies at the University of Bochum since April 2020. He studied philosophy and media studies at the Ruhr University in Bochum and the Bauhaus University in Weimar. From 2007 to 2011, Florian was a fellow at the Research Training Group Senses – Technology – Mise-en-scène (Initiativkolleg Sinne – Technik – Inszenierung), University of Vienna, and a junior fellow at the International Research Centre for Cultural Studies. Florian completed his doctorate at the University of Bochum in 2011. He was a visiting fellow at the Research Training Group Media Historiographies, Weimar and Erfurt, and Stanford University in the Department for Comparative Literature. He then went to the Digital Cultures Research Lab, at Leuphana University Lüneburg, as a postdoc via the International Graduate Center for the Study of Culture, Justus Liebig University Gießen. His scientific projects at this time focused on the history of digital cultures and smart cities. Florian was junior professor for media and cultural studies at the Institute for Theatre, Film and Media Studies at Goethe University Frankfurt for five years, working on, among other things, the biopolitics of artificial environments, media ecologies and epistemologies of the environment.

> *Writing has been my obsession ever since I learned how to hold a pen. Studying at a university meant learning how to write in different ways, how to explore ideas and problems through language. Since then, I have followed this path and I am grateful that the Johanna Quandt Young Academy has given me the time and space to finish my book.*

Current research projects

I am currently working on self-driving cars and the current transformations of traffic ('Verkehrswende' in German). In this context, I develop a conceptual and experimental framework to describe how self-driving cars interact with their surroundings by creating virtual models which operate by making micro-decisions about possible interactions with participants. To operate in an unpredictable environment, a vehicle with advanced driving assistance systems, and a robot or a drone, needs to register its surroundings and combine data from different sensors into a world model, for which it employs filter algorithms. Such world models consist of nothing other than probabilities about states and events arising in the environment. The model, thus, contains the virtuality of possible worlds that are the basis for adaptive behaviour, which includes interaction with virtual agents representing humans. My research shows that the current development of these technologies requires new concepts because their complex adaptive behaviours cannot be explained by referring them to mere algorithmic processes. Instead, I propose the heuristic instrument of micro-decisions to designate the temporality of decisions between alternatives created by probabilistic world modelling procedures. Micro-decisions are more than the implementation of deterministic processes – they decide between possibilities and, thus, always open up the potential of their otherness.

Marco Tamborini
JQYA Fellow 2020

Academy Fellowship Award

Research area: history and philosophy of science
Research focus: history and philosophy of biology from the 19th century onward, evolutionary morphology and palaeontology, techno-sciences, history and philosophy of technology

Vita

What does it mean to combine history and the philosophy of science? When Marco Tamborini asked himself this question, he was a budding philosopher in Milan and a keen reader of Kant's writings. It took him almost 10 years to arrive at a tentative and quite incomplete answer – it is about studying the broader conditions of the possibility of our knowledge, which are always locally given. To hone this interest, Marco moved for his PhD from Italy to the University of Heidelberg and afterwards to the Max Planck Institute for the History of Science in Berlin. After a post-doctorate at the Museum of Natural History in Berlin, he became assistant professor in Darmstadt, merging history and philosophy of science and technology in his habilitation. Thanks to this unusual integration, Marco has been awarded several prizes, and he has also been accepted as a visiting scholar at Clare Hall, University of Cambridge, at the Scuola Normale in Pisa, at The BioRobotics Institute | Sant'Anna School of Advanced Studies, as well as being a member of the Young Academy Mainz, which is part of the Academy of Sciences and Literature Mainz. Being part of the JQYA for Marco means being able to discuss with scholars from other disciplines in order to understand the dynamics of knowledge production.

> One of the main findings of my research is that knowledge production is always based on trans-disciplinary exchange and circulation. Innovative ideas and practices emerge from bringing together diverse fields and facilitating a productive flow of knowledge. The JQYA provides a perfect environment for cross-disciplinarily reflection. This is the reason why I enjoy working and discussing with the other JQYA fellows.

Current research projects

My current first book project, entitled The Architecture of Evolution: The Science of Form in Twentieth-Century Evolutionary Biology (manuscript under review with University of Pittsburgh Press), narrates the neglected contributions of the science of morphology to the recent development of evolutionary biology – in particular, to the field of evolutionary developmental biology (evo-devo). My book addresses the broader question about how morphological knowledge travelled, influenced other disciplines, and eventually hybridised with them. In this treatment, morphological knowledge production is tied with and rooted in different technological settings and broader philosophical frameworks. Therefore, my book reveals fruitful research traditions that have been overlooked by numerous 'classical' treatments of biology. I have published preliminary results in important international journals such as in the Journal of the History of Biology and History and Philosophy of Life Sciences. As a result, I will push the history and philosophy of a scientific discipline, evolutionary morphology, towards a broader philosophically informed and cross-disciplinarily engaged history and theory of knowledge production.

Today, more than ever, the architectural and engineering production of complex shapes is influenced by the study of organic forms. With the help of robotics, AI and 3D printers, 21st-century industry is taking inspiration from nature's organic forms and delving into the secrets of their development and function to create new materials. A continuous cycle of merging the technical and the biological is emerging, erasing the boundary between the living and the technological: we are entering an era of the 'biologisation' of technology and the technologisation of biology. How was this hybridisation possible? In a second book I am writing, I analyse recent developments in biotechnological disciplines such as bionics, biorobotics, and bio-inspired architecture. The book will be published in German and is titled, Entgrenzung: Zur Biologisierung der Technik und der Technisierung der Biologie (Dissolution of Boundaries: On the Biologization of Technology and the Technologization of Biology). Through a series of case studies and combining history and philosophy of science, I examine the premises and outcomes of this dissolution of boundaries during the 20th and 21st centuries. Preliminary results can be read in international journals like Perspectives on Science and Deutsche Zeitschrift für Philosophie.

Furthermore, I am working on a series of side projects on the history and philosophy of AI, Robotics, Bionics, and Architectural Design.

Camelia-Eliza Telteu
JQYA Fellow 2019

Academy Fellowship Award

Research area: hydrology and physical geography
Research focus: impact of climate change on freshwater systems, stakeholder-modeller dialog, hydrological modelling, science communication

Vita
Camelia-Eliza Telteu has been a postdoctoral researcher at the Institute of Physical Geography at Goethe University Frankfurt since 2016. She studied geography at the University of Bucharest and graduated with a Master's degree in Hydrology from the same university. During her doctoral studies in hydrology at the University of Bucharest, Camelia was a visiting PhD student at the University of South Bohemia, České Budějovice (Czech Republic), and at the University of Naples 'Parthenope' in Naples (Italy). In 2012, she graduated with a PhD in Geography from the University of Bucharest and became a hydrological forecaster at the National Institute of Hydrology and Water Management in Bucharest (Romania). Since 2017, Camelia has been a member of the Inter-Sectoral Impact Model Intercomparison Project (ISIMIP) that provides outcomes for the Intergovernmental Panel on Climate Change (IPCC) reports. She was a staff member of the ISIpedia – the open climate-impacts encyclopedia – between 2017 and 2020.

> I applied for a Johanna Quandt Young Academy fellowship because the academy offers me the opportunity to be creative in finding links between my research interests and JQYA themes. It encourages me to extend my comfort zone and to participate in interesting interdisciplinary debates. It stimulates me into being active and organising several workshops with JQYA fellows.

Current research projects
In 2021, Camelia continues to contribute to the ISIpedia project – the open climate-impacts encyclopedia. Her main task is to provide a harmonised description of 16 state-of-the-art global water models through a standard writing style and a standard water cycle diagram. These models contribute with simulations to the global water sector

of the Inter-Sectoral Impact Model Intercomparison Project (ISIMIP) phase 2b. She is a member of the Inter-University Planetary Thinking Group of Goethe University Frankfurt, which aims at furthering planetary thinking in the Anthropocene, thus supporting sustainability. This interdisciplinary group plans to bring together perspectives, knowledge and approaches of natural sciences, social sciences, the humanities and the arts, and citizens and stakeholders, for transformations of science and society.

Global water models (GWMs) simulate the terrestrial water cycle on a global scale and assess the impacts of climate change on freshwater systems. GWMs are developed within different modelling frameworks and consider different underlying hydrological processes, leading to varied model structures. Furthermore, the equations used to describe various processes take different forms and are generally accessible only within the individual model codes. These factors have hindered a holistic and detailed understanding of how different models operate. However, such an understanding is crucial for explaining the results of model evaluation studies, understanding inter-model differences in their simulations, and identifying areas for future model development. My latest publication provides a comprehensive overview of how 16 state-of-the-art GWMs are designed. The study analysed water storage compartments, water flows, and human water use sectors included in models that provide simulations for the Inter-Sectoral Impact Model Intercomparison Project phase 2b (ISIMIP2b). A standard writing style for the model equations was developed to enhance model intercomparison, improvement, and communication in the study. The similarities and differences found among the models analysed in this study are expected to enable scientists to reduce the uncertainty in multi-model ensembles, improve existing hydrological processes, and integrate new processes.

Members

JQYA members form the next layer of JQYA participants. They are in the process of establishing their own research group at Goethe University. Membership duration is linked to the member's independent third-party funding duration and is applicable for at least five years. Although members are not obliged to participate in the Academy's programme, we are proud to note that all members actively participate in the Academy's events with great enthusiasm. Over the past three years, we have attracted five members to Goethe University.

JQYA members form the next layer of JQYA participants. They are in the process of establishing their own research group at Goethe University. Membership duration is linked to the member's independent third-party funding duration and is applicable for at least five years. Although members are not obliged to participate in the Academy's programme, we are proud to note that all members actively participate in the Academy's events with great enthusiasm. Over the past three years, we have attracted five members to Goethe University.

Jasmin K. Hefendehl
JQYA Member 2018

Science Funding Award

Research area: neurovascular disorders
Research focus: Alzheimer's disease, ageing and comorbidity, cerebrovascular pathology, cell function in the neurovascular unit, two-photon microscopy

Vita
I studied biology with a focus on neuroscience at Ulm University from 2003 to 2007, after which I worked at University College London as a student assistant for nine months to expand my methodological spectrum. In 2008, I started my PhD with a

stipend award at the Hertie Institute for Clinical Brain Research at the University of Tübingen, where I investigated Alzheimer's disease pathology as well as the immune system of the central nervous system. Upon completing my PhD in 2012, I stayed at the Hertie Institute for Clinical Brain Research for a short first postdoctoral position to work on the ageing of microglial cells, which are the brain's primary immune cells. In 2012, I received a DFG (German Research Foundation) postdoctoral stipend to conduct research on early cell dysfunction in Alzheimer's disease and the role of the neurovascular unit in stroke recovery at the University of British Columbia at the Djavad Mowafaghian Centre for Brain Health. Whilst abroad, I received further funding from the DFG and the Michael Smith Foundation for Health Research. Upon my return to Germany in 2016, I received a stipend from the DFG to reintegrate into the German system and apply for an Emmy Noether Programme granted in 2018. Since then, I have been heading a research group at Goethe University, which focuses on the co-morbid state of Alzheimer's disease and vascular cognitive impairments such as stroke.

> I believe that the complexity of modern science can only be addressed and understood as a team, providing several layers of expertise toward the understanding of a grand theme. This not only includes researchers from the same scientific area but an interdisciplinary cooperation, providing novel intellectual approaches which otherwise would not be accessible.

Current research projects

Alzheimer's disease (AD) is a progressive neurodegenerative disorder that remains one of the significant burdens on our society with no available cure. In combination with vascular cognitive impairment (VCI), AD represents over 80% of the total dementia cases. At autopsy, the overlap of the underlying pathological hallmarks of AD (beta-amyloid (Ab) plaques and neurofibrillary tangles) and VCI (strokes, small vessel disease and vascular dementia) is pronounced but often unrecognised in the clinical setting. Hence, we are focused on the basic research of the comorbidity states of AD and VCI to investigate underlying disease pathways as well as potential biomarker profiles. We use various state of the art techniques, such as 2-Photon imaging, RNA sequencing and primary cell culture models of the blood-brain barrier, to investigate pathological alterations of the comorbid states. The neurovascular unit comprises different cell types that work together to ensure vital functions such as blood-brain barrier integrity, regulation of cerebral blood flow and immunoregulation. Pericytes are part of the neurovascular unit and are of particular interest to us, as both pathologies impact it and hence creates a common link between the diseases. Collaboration with clinicians enables us to work in a translational setting and validate our human samples' findings. The overall goal is to investigate the long-term interplay of the two pathologies and identify new molecular mechanisms leading to alterations of the neurovascular unit and potential new disease markers. The JQYA funds have enabled me to acquire the necessary lab equipment for our research and start several pilot projects to explore this timely research question. We have, for example, found that microglial cells, the primary immune cells of the brain, form a new and yet uncharacterised sub-population in the comorbid state, which we are currently investigating with the use of the JQYA funds. These hold tremendous implications in terms of tissue recovery after stroke and long-term changes to the underlying Alzheimer's disease pathology.

Benesh Joseph
JQYA Member 2018

Science Funding Award

Research area: biophysics
Research focus: membrane transport mechanism, membrane biogenesis and in-situ / cellular structural biology

> *My transition towards a scientific career was a rather spontaneous outcome of my intrinsic na-ture of questioning things while trying to understand them. I am very thankful that a scientific career allows me to deeply explore such curiosity-driven questions with outputs beneficial for humanity. Being a part of the JQYA has been very crucial for establishing my research group at Goethe University.*

Vita

Benesh Joseph started his academic career at Osaka University Japan where he studied biotechnology first, with a UNESCO-JAPAN Fellowship in Biotechnology followed by a Monbukagakusho Fellowship. In 2008 Benesh graduated with a Master of Engineering from Osaka University. Between 2013 and 2016, he conducted his postdoctoral research at Goethe University Frankfurt and as a visiting researcher at the University of Virginia, USA. He was awarded with a Marie-Curie Postdoctoral Fellowship from 2013 to 2015, with an Adolf-Messer Prize for Young Investigators in 2017. Since 2016, Benesh has been an independent investigator at Goethe University Frankfurt.

Christian Münch
JQYA Member 2018

Science Funding Award

Research area: biochemistry
Research focus: protein quality control, protein folding and responses to mis-folding, stress responses in mitochondria, cancer and neurodegeneration

I have always been fascinated by understanding how life works at a molecular level. This curiosity drove me to study biochemistry and then to become a scientist to provide new explanations and discoveries myself.

Vita

Christian Münch has been an Emmy Noether Programme group leader at the Gustav-Embden Centre of Biochemistry, Goethe University Frankfurt since 2016. Christian studied biochemistry at the University of Tübingen and the Max Planck Institutes in Martinsried and Tübingen. He obtained his PhD from the University of Cambridge (UK) and earned the prestigious British Neuroscience Association Postgraduate Award. After his graduation, Christian joined the Harvard Medical School with an EMBO long-term fellowship to study protein quality control in mitochondria. His current research continues on mitochondrial quality control, determining the processes involved in the mitochondrial unfolded protein response and its role in human diseases. Christian was awarded an ERC starting grant in 2018 and an Aventis Foundation Postdoctoral Award in 2019.

Cornelia Pokalyuk
JQYA Member 2020

Science Funding Award

Research area: mathematics, specialisation in probability theory, biomathematics
Research focus: stochastic dynamics of host parasite processes, population genetic processes with spatial structure and selection, coalescing and branching processes

Vita
Cornelia Pokalyuk is a mathematician with pronounced interests in biology and medicine. Since autumn 2020, she has been leading an Emmy Noether group on 'Stochastic dynamics of parasite evolution' at Goethe University Frankfurt. Cornelia completed her diploma in mathematics with a minor in computer science at the University of Leipzig. During her PhD at the University of Freiburg, she was a member of the DFG-funded research group on 'Natural selection in structured populations'. As a doctoral student of Prof. Pfaffelhuber, she studied population genetic processes with spatial structure and selection. To learn about current research topics in biology and medicine, she joined the group of Prof. Jensen at the School of Life Sciences at École Polytechnique Fédérale de Lausanne and investigated within-host infection histories of the herpesvirus HCMV on the basis of whole-genome samples. Returning to mathematics, she analysed stochastic models that capture the specific evolutionary dynamics of HCMV populations as a post-doctorate in Prof. Wakolbinger's group at Goethe

University. From autumn 2015 until autumn 2017 Cornelia first held a Dorothea Erxleben guest professorship and then an interim professorship at Otto von Guericke University Magdeburg, and in the summer term 2020 she was an interim professor at Justus Liebig University Gießen.

> *Since my childhood, I have loved mathematics, and I am intrigued with medicine. My passion is the research at the interface of both disciplines! At JQYA, I enjoy the interdisciplinary exchange. It provides fascinating insights into research areas that I only rarely get in touch with!*

Current research projects

Cornelia Pokalyuk's research focuses on the mathematical analysis of population dynamics, particularly of parasite populations. She is especially interested in the interplay between random fluctuations and deterministic dynamics. Stochastic effects play a role, for instance, when relatively small populations are involved, as is the case when parasites invade a (so far unaffected) host population, or an advantageous variant arises in a population. Furthermore, the analysis of stochastic fluctuations is fundamental for statistical data analysis.

The establishment or fixation probability of an advantageous variant is a cornerstone in population genetics and the basis for analysing more complex scenarios, like adaptation to a changing environment or emergence of drug resistance. Deterministic as well as random drivers govern establishment probabilities. While it is increasing in the average offspring numbers of carriers of the advantageous variant, it is decreasing in the offspring variance. Cornelia has developed, together with her co-authors, several methods for the analysis of these probabilities in various scenarios. These methods partially rely on a forward and partially on a backward perspective of the fixation process and can often be utilised to identify the corresponding times to establishment or fixation.

With respect to parasite evolution, Cornelia is particularly interested in the analysis of mechanisms that allow for the coexistence of parasite and host. In close exchange with experimental groups, she develops individual-based models that map observed host-parasite dynamics. Parasite evolution is substantially influenced by the dependence of the parasites of their host. For instance, the spread of the parasite population over the host population can imply a hierarchical structure of the parasite population, and the spatial structure of the host population can carry over to the parasite population. With the help of genealogical processes and separation of scales both in time and space, the proposed models are analysed so that the driving dynamics of the process become accessible.

Andreas Schlundt
JQYA Member 2019

Science Funding Award

Research area: structural biology
Research focus: mRNA-based gene regulation, RNA-intrinsic elements, regulatory proteins, integrated structural biology approach, NMR spectroscopy, cancer and autoimmunity, RNA viruses

Vita

Andreas Schlundt was born in Berlin, where he went to school, interrupted by a one-year exchange stay in Russia. He then studied biochemistry at the Free University Berlin. After an internship at the Swiss Federal Institute of Technology (ETH) Zurich, Andreas graduated in 2007 with a diploma from the Leibniz Institute for Molecular Pharmacology in Berlin. In the same Institute and at the Free University Berlin, Andreas conducted his doctoral studies on the structural biochemistry of the immune system and graduated with the highest distinction in 2012. Andreas moved on for his postdoctoral research to the Technical University Munich and Helmholtz Centre Munich, where he continued working in the field of molecular structures, dynamics and interactions of biological macromolecules. This particular stay brought him in contact with the world of protein-RNA interactions in their broad flavours, which he studied with his core method of Nuclear Magnetic Resonance (NMR) Spectroscopy. Since 2018, Andreas has been carrying out his research as an independent group leader of the DFG-funded Emmy Noether Programme at Goethe University Frankfurt, where he joined the Institute of Molecular Biosciences and the Biomolecular Magnetic Resonance Centre. In 2020, he became one of the founders, core organizers and principal investigators of the Frankfurt-based, international Covid19-NMR consortium, actively contributing to global campaigns for fighting the pandemic.

> I was fascinated very early in my academic career by the idea of understanding how life works at the molecular level. I am still driven by the desire to visualize the invisible in fundamental biological processes looking at molecules doing their precise work for a cell, a tissue and an entire organism.

Current research projects

We focus on a particular stage of gene regulation: the level of messenger RNA (m) RNA, positioned between the DNA as a storage place of genetic information and the gene products. mRNA 'fates', like their transport through cells, modifications, stability and turnover, are strictly controlled by RNA-binding proteins (RBPs), thereby steering the abundance of a gene product in time, space and quantity. A misregulation of mRNAs is a critical parameter for, for example, autoimmune diseases, tumour progression, and after viral infections.

RNAs occur in various species regarding their shapes, folds, sequence composition and dynamic behaviour. Consequently, RBPs contain different types of domains that are capable of interacting with particular RNA motifs. We are interested in the various domains and their appearance in combinations within RBPs in order to understand the complex networks with RNAs. Our ongoing projects focus on studying the immunoregulatory protein Roquin, the oncofetal protein IGF2BP3, the pro-inflammatory protein Arid5a and other proteins involved in RNA processing within human cells. Our work has taken a special route towards coronaviral RNAs interacting with coronaviral and host (human) proteins. RNA viruses, like Coronaviruses, bear prime examples of the importance of RNA-protein interactions for an intact life cycle. We mainly focus on the so-called nucleocapsid protein, which serves the virus to package its genome for progression and spread.

Our methodological approach centres on the determination of highly resolved structures, i.e. atomistic pictures of those RNA-protein complexes, following advanced sample preparations at high amounts, stability and homogeneity. We focus on solution-based methods, most of all NMR spectroscopy and X-ray scattering, keeping the molecules of interest in a native-like state and allowing us to analyse the role of their intrinsic dynamics within the biological system. In particular, in terms of the challenging financial demands of our methods, the JQYA Science Funding facilitates our progress. It allows the non-bureaucratic, non-complicated purchase of technical lab devices and covers the costs for advanced sample shipments to X-ray beam time, software, external services. My lab would and could not be where and what it is without this highly appreciated financial support.

Distinguished Senior Scientists

Distinguished Senior Scientists build an external board of the JQYA. These are renowned international researchers with extraordinary scientific achievements and unprecedented commitment to young scientists. They play a crucial role by giving support and advice to the Academy's directors and by delivering critical and inspiring impetus to the fellows' scientific discussions.

Distinguished Senior Scientists build an external board of the JQYA. These are renowned international researchers with extraordinary scientific achievements and unprecedented commitment to young scientists. They play a crucial role by giving support and advice to the Academy's directors and by delivering critical and inspiring impetus to the fellows' scientific discussions.

Seyla Benhabib
Distinguished Senior Scientist 2018–2021

Humanities

Seyla Benhabib is a Turkish-American philosopher and the Eugene Meyer Professor of Political Science and Philosophy Emerita at Yale University, where she was director of the program in Ethics, Politics and Economics (2002–2008). Professor Benhabib is internationally known for her work in political philosophy, which draws on critical theory and feminist political theory, and has written extensively on Hannah Arendt and Jürgen Habermas and the topic of human migration.

Professor Benhabib was elected to be the corresponding fellow of the British Academy for the Humanities and Social Sciences (summer 2018). She was president of the Eastern Division of the American Philosophical Association in 2006–2007, a fellow at the Wissenschaftskolleg in Berlin (2009), at the NYU Straus Institute for the Advanced Study of Law and Justice (Spring 2012), and the German Marshall Fund's Transatlantic Academy in Washington DC (Spring 2013). In 2009, she received the

Ernst Bloch Prize for her contributions to cultural dialogue in a global civilization and, in 2012, the Leopold Lucas Prize from the Evangelical Academy of Tübingen. She holds honorary degrees from the Humanistic University in Utrecht (2004), the University of Valencia (2010), Boğaziçi University (2012), Georgetown (2014), Université de Genève (2017) and Centro de Estudios Latinoamericanos de Chile (2021). She received a Guggenheim grant in 2010 and 2011 for her work on sovereignty and international law. Professor Benhabib was awarded the Meister Eckhart Prize from the Identity Foundation and the University of Cologne in 2014 for her contribution to contemporary thought. She is a senior research fellow at Columbia Law School and the Center for Contemporary Critical Theory at Columbia University and was a scholar in residence at the Columbia Law School (2018–2019), where in spring 2019 she held the chair of James S. Carpentier Visiting Professor of Law. She has been a member of the American Academy of Arts and Science since 1996. Professor Benhabib held the Gauss Lectures (Princeton, 1998); the Spinoza Chair for distinguished visitors (Amsterdam, 2001); the John Seeley Memorial Lectures (Cambridge, 2002), the Tanner Lectures (Berkeley, 2004) and was the Catedra Ferrater Mora Distinguished Professor in Girona, Spain (summer 2005).

Professor Benhabib's most influential books include 'The Rights of Others: Aliens, Citizens and Residents', Cambridge University Press, 2004 (German: 'Die Rechte der Anderen', Suhrkamp, 2008); 'Hannah Arendt: Die Melancholische Denkerin der Moderne', Rowohlt, Hamburg, 1998, neue Ausgabe, Suhrkamp, 2006; 'Kosmopolitismus ohne Illusionen: Menschenrechte in unruhigen Zeiten', Suhrkamp 2016. She has edited eight volumes, ranging from discussions of communicative ethics to democracy and difference to identities, allegiances and affinities. 'Migrations and Mobilities. Gender, Borders and Citizenship', edited with Judith Resnik of the Yale Law School (NYU Press 2009), was named a 'Choice outstanding book'. 'Toward New Democratic Imaginaries: Istanbul Dialogues on Islam, Culture and Politics', edited by Seyla Benhabib and Volker Kaul (Springer Verlag, November 2016), was based on seminars conducted by Professor Benhabib and the RESET Foundation at BILGI University in Istanbul from 2008 to 2015.

Over her long career, Professor Benhabib has supervised close to 50 dissertations in the United States and has sponsored international young scientists from Germany, Spain, Italy, Turkey, Israel and Brazil.

Gunnar von Heijne
Distinguished Senior Scientist 2018–2021

Natural and Life Sciences

Professor Gunnar von Heijne is a Swedish scientist working on signal peptides, membrane proteins, and bioinformatics in biochemistry and biophysics at Stockholm University. He graduated in 1975 with a Master of Science degree in Chemical Engineering from the Royal Institute of Technology (KTH). He then became a doctoral student in theoretical physics at KTH, a research group focusing on statistical mechanics and theoretical biophysics, and was awarded his PhD in 1980. In 1983 he was made a docent in theoretical biophysics at KTH, where he remained until 1988. He was active as a science reporter at the Swedish National Public Radio from 1982 to 1985. From 1989 to 1994 he was an Associate Professor at Karolinska Institute, and in 1994 he was made a professor in theoretical chemistry at Stockholm University.

Professor von Heijne has received honorary doctorates from Åbo Akademi (2008) and the University of Valencia (2019). He has been awarded the van Deenen Medal from Utrecht University (2009), the Accomplishment by a Senior Scientist Award by the International Society for Computational Biology (2012), the Novo Nordisk Foundation Novozymes Prize (2018), and the Biophysical Society Anatrace Membrane Protein Award (2019).

Professor von Heijne has been a member of the Royal Swedish Academy of Sciences since 1997. He was also a member of the Nobel Committee for Chemistry from 2001 to 2009 (chairman 2007–2009) and served as its secretary from 2014 to 2020. He was a member of the Board of the Nobel Foundation between 2015 and 2021. His Nobel Prize engagements included various activities in science communication and interactions with young scientists across the world.

Professor von Heijne's research primarily concerns membrane proteins, and he is one of the most cited Swedish scientists in diverse biochemistry and molecular biology areas. He has been director of Stockholm Bioinformatics Center and the Center for Biomembrane Research at Stockholm University and the SciLifeLab National Cryo-EM Facility in Stockholm. He was vice director of SciLifeLab from 2009 to 2015.

Christiane Nüsslein-Volhard
Distinguished Senior Scientist 2021–2024

Natural and Life Sciences

Christiane Nüsslein-Volhard is a German biologist and biochemist. She studied biology, physics and chemistry at Goethe University Frankfurt and obtained her diploma in biochemistry in Tübingen in 1968. She received her doctorate in natural sciences from the University of Tübingen (1973) in genetics.

Christiane Nüsslein-Volhard's research focuses primarily on questions of evolutionary biology, in particular on forms and shapes during the development of animals. From 1985 to 2014, she was director of the Department of Genetics at the Max Planck Institute for Developmental Biology in Tübingen. Since then, she has been leading a powerful emeritus research group at the same MPI entitled 'Colour pattern formation'.

Christiane Nüsslein-Volhard has received numerous awards, honorary doctorates and prizes for her discoveries, including the Leibnitz Prize of the German Research Foundation (1986), the Albert Lasker Medical Research Award (1991) and, as the first German woman ever, the Nobel Prize for Medicine (1995). She is a member of the Royal Society (England), the National Academy (USA), the Leopoldina (Germany), the Berlin-Brandenburg Academy (Germany), the Curia of Science (Austria) and the Académie des Sciences (France). As a member of the German Ethics Council (2001–2006), Christiane Nüsslein-Volhard has always faced the critical questions of genetic engineering and discussed them openly. She was president of the Society of German Natural Scientists and Physicians (until 2008) and secretary-general of the European Molecular Biology Organisation (EMBO) (until 2009). Since 2005 she has been a member of the Scientific Council of the European Research Council (ERC) of the European Union, and since 2013 she has been chancellor of the Order Pour Le Mérite for Sciences and Arts, of which she has been a member since 1997. Christiane Nüsslein-Volhard has published around 200 publications in scientific journals, numerous newspaper articles (FAZ, Die Zeit) on current topics, as well as several books for general readers: 'Das Werden des Lebens' published by CH Beck, 'Von Genen und Embryonen' published by Reclam and 'Schönheit der Tiere' published by Matthes & Seitz.

For many years, Christiane Nüsslein-Volhard has been committed to promoting women scientists. The Christiane Nüsslein-Volhard Foundation for Women in Science, which she founded in 2004 and where she is still active as chairperson, supports talented young female scientists with children to give them the freedom and financial support they need for a scientific career. Furthermore, Christiane Nüsslein-Volhard mentors young female and male researchers on their way to a scientific career of their own. In summer 2020, she was appointed honorary senator of the Max Planck Society. With this award, the members of the General Assembly of the Max Planck Society honoured her decades of top-level research and highlighted her extraordinary commitment to promoting female scientists.

Hartmut Rosa
Distinguished Senior Scientist 2021–2024

Social Sciences

Hartmut Rosa is a German sociologist and political scientist. He studied philosophy, political science and German studies in Freiburg and London from 1986 to 1993. He then received a doctoral scholarship from the German National Academic Foundation (1993–1996). With his dissertation 'Identity and Cultural Practice. Political Philosophy after Charles Taylor', he was awarded a doctorate summa cum laude in 1997 at the Humboldt University in Berlin. In 2004 he habilitated with a thesis on 'Soziale Beschleunigung. Die Veränderung der Zeitstrukturen in der Moderne' (Social Acceleration: A New Theory of Modernity) in the subjects of sociology and political science at the University of Jena, where he is still active today as professor of General and Theoretical Sociology.

Hartmut Rosa is a specialist in the topics of time diagnosis and modernity analysis, normative and empirical foundations of social criticism, subject and identity theories, sociology of time and acceleration theory, and currently with what he calls the 'sociology of world relations'. Particularly noteworthy are his analyses of the thematic spectrum of 'resonance' and the question of 'unavailability' as attitudes or attitudes of people in our culture. Rosa has received several awards for his research and exceptional services to science, including the Tractatus Prize for Philosophical Essay Writing (2016) and the Erich Fromm Prize from the International Erich Fromm Society for the Recovery of Humanistic Thought and Action in the Sense of Erich Fromm (2018). In 2020, he was awarded the Werner Heisenberg Medal by the Alexander von Humboldt Foundation to recognise his services to international scientific cooperation and his many years of work on the Anneliese Maier Selection Committee.

As director of the Max Weber Kolleg for Cultural and Social Studies at the University of Erfurt, Hartmut Rosa has dedicated many years to promoting diversity and equal participation and transparency in theory and practice in a prejudice-free working environment. Rosa's research contributes significantly to the 'education of global citizens', thus educating them to become responsible individuals in a globalised world. For this, he received the Rob Rhoads Global Citizenship Education Award from the University of California, Los Angeles, in 2020. The fellows of the Max Weber Kolleg – like the JQYA fellows – work on an interdisciplinary research project for a limited period. Furthermore, the Kolleg stands for intergenerationality, internationality and intersectorality.

His contribution to promoting interdisciplinarity and internationalisation in research was recognised at the Salzburg University Week, where he held a keynote lecture at the closing ceremony. The Salzburg University Weeks are an annual summer university of the University of Salzburg. The aim is to form a university forum where theology and all other sciences address both fundamental and current questions and problems of our time. Hartmut Rosa feels committed to the freedom and independence of research and teaching – two crucial principles of the JQYA – and shows solidarity with scholars worldwide who do not enjoy this freedom.

Nicola Spaldin
Distinguished Senior Scientist 2018–2021

Natural and Life Sciences

Nicola Spaldin is a professor of materials theory at ETH Zurich, known for her pioneering research on multiferroics. Professor Spaldin graduated from the University of Cambridge in 1991 with a Bachelor's degree in Science and earned her PhD in Chemistry from the University of California, Berkeley in 1996. The thesis topic was 'Calculating the Electronic Properties of Semiconductor Nanostructures: A New Theoretical Approach'. As a postdoctoral fellow, she worked with K. M. Rabe at Yale University before obtaining her first professorship (assistant professor) at the University of California, Santa Barbara, in 1997. In 2002, she became an associate professor, and in 2006 she received a full professorship. She moved to the ETH in 2011.

Professor Spaldin uses theoretical and computational techniques to develop and explain new materials with novel and potentially technologically relevant properties. She is particularly well known for creating a new class of crystalline compounds, called multiferroics, that are simultaneously ferromagnetic and ferroelectric. She theoretically justified why there are so few multiferroics, publishing her explanation in a landmark paper in 2000, and proposed suitable materials for experimental synthesis based on her computational studies. Following her theoretical predictions, she was part of a team that experimentally demonstrated the multiferroic properties of the most widely studied multiferroic, bismuth ferrite, in particular that its magnetic properties could be modified using an electric field. Her ideas brought about a worldwide revival of multiferroics research, which continues to be active today. Professor Spaldin was the winner of the American Physical Society's James C. McGroddy Prize for New Materials 2010. In 2015, she won the Körber European Science Prize for 'laying the theoretical foundation for the new family of multiferroic materials', and, in 2017, she was the European laureate of the 2017 L'Oréal-UNESCO Awards for Women in Science. Two years later, she was awarded the Swiss Science Prize Marcel Benoist, which is given to Swiss nationality or residency scientists for the most valuable scientific discovery of the year. Professor Spaldin is a fellow of the American Physical Society (2008), the Materials Research Society (2011), the American Association for the Advancement of Science (2013), the Royal Society (2017) and foreign member of the National Academy of Engineering (2019).

In addition to her research interests, Professor Spaldin is a passionate and effective educator, author of a popular textbook on magnetic materials, winner of multiple awards for teaching excellence, and coordinator of her department's curriculum development program 'The Materials Scientist 2030, Who is She?'. When not in the laboratory, she can usually be found playing her clarinet, skiing or climbing in the Alps.

Eleonore Stump
Distinguished Senior Scientist 2021–2024

Humanities

Eleonore Stump is the Robert J. Henle Professor at Saint Louis University. She holds a B. A. in Classical Languages from Grinnell College, where she was valedictorian (1969). She also holds an M. A. in New Testament Studies from Harvard University (1971) and an M. A. and PhD in Medieval Studies from Cornell University (1975). Before moving to Saint Louis University in 1992, she taught at Oberlin College, Virginia Tech, and Notre Dame. She has also had visiting appointments at Goethe University (Frankfurt), Calvin College, Oxford (Oriel), Baylor University, the Pontifical Gregorian University, Aberdeen University, Princeton University, Wuhan University, and the Australian Catholic University.

Among her many distinguished lectures, Eleonore Stump has given the Gifford Lectures (Aberdeen), the Wilde Lectures (Oxford), the Stewart Lectures (Princeton), and the Stanton Lectures (Cambridge). Her numerous awards include the Robert Foster Cherry Award for Great Teaching (Baylor University, 2004) and honorary doctorates from Marquette University (2006), Tilburg University ((2017), and Austal University in Argentina (2021). She received the Aquinas Medal from the American Catholic Philosophical Association (2013) and a Lifetime Achievement Award from the American Maritain Association (2017). In 2012, for her life's work – outstanding teaching, publications of lasting scholarly value and influence on philosophical thought – Eleonore Stump was inducted into the American Academy of Arts and Sciences. In 2014 to 2015, with her colleague John Greco, she was awarded a $ 3.3 million Templeton grant for work on the topic 'Intellectual Humility'. She has been president of the Society of Christian Philosophers (1995–1998), the American Catholic Philosophical Association (1999–2000), the American Philosophical Association, Central Division (2005–2006), and Philosophers in Jesuit Education (2013–2015).